Alison Butler

Critical acclaim for this book

'A unique and extremely valuable book. With innovative and clear graphic presentations of information, Sutcliffe provides an enlightening view of the inequalities that plague the world economy. Each image is accompanied by a brief explanation and a short, effective commentary. It is the graphs themselves, however, that are the heart of the book. They are marvellous tools for provoking discussion of complex economic issues, and thus they are exceptionally useful in the classroom. It is a new proof of an old adage: a picture is worth 1,000 words!'

ARTHUR MACEWAN, PROFESSOR OF ECONOMICS, UNIVERSITY OF MASSACHUSETTS

'These vivid images of the vast and complex inequalities of our world catch the eye and tear the heart. They illustrate a powerful economic analysis of the many forms of poverty and oppression that emerge from conflicts based on nation, class, and gender.'

NANCY FOLBRE, PROFESSOR OF ECONOMICS, UNIVERSITY OF MASSACHUSETTS, COAUTHOR OF *THE ULTIMATE FIELD GUIDE TO THE AMERICAN ECONOMY*

About the author

BOB SUTCLIFFE is an economist who was educated at Oxford and Harvard universities. During his long career, he has taught at many different universities in Britain, the United States of America, Nicaragua and, most recently, at the University of the Basque Country in Bilbao, where he works with Hegoa, an institute for the study of development and the international economy. He is the author of several books, including *Industry and Underdevelopment*.

100 ways of seeing
an unequal world

BOB SUTCLIFFE

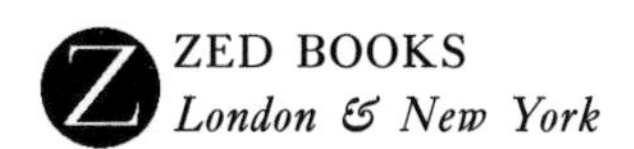
ZED BOOKS
London & New York

100 Ways of Seeing an Unequal World
was first published in an updated revised edition by Zed Books Ltd,
7 Cynthia Street, London, N1 9JF, UK and Room 400, 175 Fifth Avenue,
New York NY 10010, USA in 2001.

Published originally in Spanish as *100 imágenes de un mundo desigual*
by Intermón Fundacion para el Tercer Mundo,
Apartado no. 310 F.D., 08080, Barcelona, in 1998.

Distributed in the United States exclusively by Palgrave,
a division of St Martin's Press, LLC, 175 Fifth Avenue, New York,
NY 10010, USA.

Second impression 2002

Cover design by Andrew Corbett.
Typeset in Monotype Ehrhardt by Illuminati, Grosmont.
Printed and bound by Biddles Ltd, *www.biddles.co.uk*

A catalogue record of this book is available from the British Library

US CIP data is available from the Library of Congress

ISBN 1 85649 813 1 hb
ISBN 1 85649 814 X pb

Contents

Introduction
Our unequal world

Dos linajes solos hay en el mundo, como decía una abuela mía, que son el tener y el no tener.

[There are only two families in the world, as a grandmother of mine used to say: the haves and the have-nots.]

Observation of Sancho Panza to Don Quijote, in Miguel de Cervantes Saavedra, *Don Quijote de la Mancha*

The purpose of the book

This book is, first of all, designed to communicate, by primarily visual means, a large amount of information about inequalities in human society. Using graphs and maps it tries to follow some of the (demanding) precepts about the visual display of information proposed by Edward Tufte in his absorbing and beautiful books on the subject (1). The original Spanish edition arose out of a proposal from my friend the late Ignacio Senillosa of Intermon in Barcelona to update a book by Rudolf Strahm (2) which uses the same technique of alternating visual and verbal material. In that book, as in this, it is the pictures which are most important.

I hope that this approach will make the book a useful tool in a variety of educational contexts. It is designed for university and adult education courses about development, inequality, international economics and international relations, although it contains material relevant to other fields. All the material is organized around the central theme of inequality, or rather inequalities based on class, gender, ethnicity, country, place of residence, sexual orientation, and political power as well as other factors.

The book does not contain an explicit general argument about inequality. Its intention is rather to present material which will provoke, both inside and outside lecture rooms, thought, discussion and argument about the causes, consequences and possible remedies for inequality. A picture should save words but can also generate words. Those words should come from those of you who use the book. You will find, however, that it contains an implicit but obvious egalitarian standpoint. The inequalities that are described here are, in my view, evidence for the existence of profound unacceptable social injustices; the reason for measuring and discussing the inequalities is to help to find ways to remedy the injustices.

I believe that it is important for socialists today to recapture the egalitarian spirit which was the heart of the historical utopian socialisms, including Marxism (3). Equality in this tradition is very far from homogeneity or conformity; it means the protection of freedom through equal civil and political rights and distribution according to need and not according to the ability to pay.

Debates about inequality

Questions of the relationship of inequality with injustice, with freedom or with efficiency have always been central to philosophical, political and economic debates. The present revival of interest in them is due to several factors: increasing scepticism about whether a continuously rising economic tide will lift all boats, new social movements aiming to redress specific inequalities, new theoretical formulations of egalitarian principles by social philosophers, a growing interest over a wide range of the political spectrum in the notion of human rights, the accumulation of ever greater amounts of empirical information allowing inequalities to be quantified and, most recently, revulsion at some of the consequences of the economic liberalization practised throughout the world since Margaret Thatcher and Ronald Reagan led their crusade to let self-interest off the leash.

Those politicians and their supporters were consciously implementing the ideas about equality and inequality which had been propounded for centuries by uncritically conservative economic thinkers – those who believed in the justice of market capitalism and the efficacy of the famous 'invisible hand'. These orthodox or neo-liberal economists do not advocate inequality as such. They argue that a market-driven economy will, by stimulating risk and enterprise, constantly create new inequalities as some people prosper from new activities, but will just as constantly eliminate inequalities as the new activities spread via competition and the market. The distribution produced by the market and the just distribution are to them one and the same thing. All in all, over the long run, they predict that global market capitalism can successfully develop the whole world and that free competition will equalize economic rewards between countries if not between individuals. They seem to foresee a kind of 'equality of inequalities'.

The more consistent advocates of the free market have an explanation as to why historically even this result seems to be so long in coming: it is that the market is not free enough; monopoly and 'imperfections' hamper the mechanism. But most of them do not bother too much with that question. What does bother all of them, however, is the related idea that the deliberate pursuit of redistribution is dangerous and perverse: it will both kill the potential benefits of the market by lowering growth and may even go so far as to worsen the position of those it is designed to help. An ebbing tide will sink all boats. In this curious sense, therefore, orthodox economics, while not explicitly against equality, is emphatically anti-egalitarian. Any attempt to redivide the cake, it says, will make it shrink. The poor, it asserts, will be better off with a small slice of a large cake than a large slice of a small cake. There is, in other words which are often used, a trade-off between equity and efficiency.

Critics of inequality must recognize that the size of the cake (the total amount of production) is not independent of the size of the slices of it which people are expected to divide it into. Most economic activities are undertaken because they are expected to produce certain benefits for their initiators. If they cannot realize these benefits they will not undertake the activity. The economic cake has no given size, allowing it to be sliced up by political decisions in any way that the people or its government decides. Many reformists have seemed to have a naive view that it does.

This point, however, is not necessarily an argument against egalitarian measures, which is how the orthodox economists use it. There is considerable evidence to support a diametrically opposed proposition: that it is the inequality of the slices into which the cake is cut that restricts it size. Inequality and not equality impedes prosperity and growth. The history and present of economics contains many arguments of this kind, although their authors have often been considered suspiciously heterodox.

The well-known heterodox economist J.A. Hobson, author of the much-quoted book *Imperialism: A Study*, written in 1902, argued that British imperialism in Africa, which he deplored, was the outcome of inequality in the distribution of income in Britain: because the purchasing power of a large part of the population was restricted, a lower rate of profit forced capitalists to seek profits abroad and they needed imperialist rule to protect them. Measures to equalize distribution in Britain would, he believed, both be desirable in themselves and produce employment and growth without imperialism. Keynes later used similar arguments to say that inequality in income distribution reduced the level of aggregate demand and so made the economy more vulnerable to slump and unemployment. Keynesian economic historians have rather convincingly blamed at least a part of the Great Depression which battered the world from 1929 onwards on the rise of inequality which accompanied the United States' boom of the 1920s. Growing inequality eventually stifled demand for new industrial products and the economy collapsed. A recent collection of studies on the consequences of inequality concluded that 'the aggregate evidence does not support the claim for a positive relationship between inequality and efficiency' (4). These studies, mostly relating to developed countries, stress the cost to society of not using human resources as fully as possible, and so of curtailing the amount of goods and welfare available. Another argument, applied to developing countries, is that the main explanation of the relative development success of East Asian countries compared with the relative failure of Latin American ones is the fact that income distribution in the latter is much more unequal than in the former. It is argued that less inequality in Asia makes for much larger internal markets for non-basic goods and services in Asian countries, while Latin American inequality only permits a small market for luxuries. Equality is thus said to be better for development (5). There are good reasons to think that it is also an important condition for sustainability in development (6). It is also argued that equality may help development and many other desirable goals by reducing social conflict. This is part of a more general argument that inequality entails enormous policing costs. Unequal societies use a great quantity of resources to maintain the inequality: more prisons and prison guards, more police, more private security services to protect property and life, more expenditure on weapons and on security goods such as locks, along with factory, shop, car and house security systems. These activities, encouraged by inequalities, must occupy a large proportion of the labour force.

All these arguments deny the existence of a trade-off between equity and economic efficiency; the two are mutually reinforcing. The variety and persuasiveness of such arguments is impressive. But they have a collective weakness. They are empirical arguments which advocate equality because it produces good results. But empirical relationships have a nasty habit of being refuted by further evidence or of reversing themselves with time. I would be the last person to say that the empirical evidence is not important; but it is in the end insufficient as a justification for egalitarianism. Redistribution is desirable not only if it increases economic efficiency. Equality is desirable independently of its consequences because it is the central component of social justice, independent of its consequences. The pursuit of justice is an end in itself.

The publication in 1971 of the philosopher John Rawls's *The Theory of Justice*, combined with the simultaneous rise of great social and political movements in favour of more equality such as feminism and the civil rights movement, gave new life to debates about justice and equality. Rawls's suggestion that social justice demanded a 'maximin' approach in which the position of the worst off was maximized has proved to be controversial, but has been a stimulant to further ideas about equality, from Michael Walzer's search for a definition of 'complex equality' which more clearly distinguishes it from identity, to Amartya Sen's insistence that equality of outcomes is less fundamental than equality of capabilities, an idea related both to human development (graph 21) and to more rigorous explorations of concepts like equality of opportunity. None of these contributions makes a case for exact material equality between individuals; but in various ways they strengthen the case against the kind of inequalities seen in this book and in favour of active egalitarian policies to improve social justice. This is important at a moment when human society appears in so many ways to be moving in the opposite direction.

Is inequality growing?

World inequality is not a simple concept that can be assigned a single value. Multitudinous inequalities exist in human society – of income, welfare, rights, power and prestige, based on class, caste, gender, ethnicity, age, sexual preference and many more things. A particular individual can simultaneously gain from one form of inequality and lose from another. In no period of human history have all existing inequalities become greater or less. In fact the same event often both reduces and increases inequality at the same time. If there are three people, A, B and C, who receive respectively 10, 5 and 1 units of something desirable, and the situation then changes and B starts to receive 7 units instead of 5, what has happened to inequality? The inequality between A and C is unchanged, that between A and B is reduced and that between B and C is increased. The world is full of such cases and if we want to say whether inequality in general has increased we must give a very precise definition of a complex concept.

The widespread feeling that we live in an epoch of rapidly rising inequality is constantly fuelled by information from press or television or from simple observation in the street. The very rich seem richer than ever; furthermore, stock market booms and the 'new economy' have swelled their ranks (graph 12). Meanwhile at the other extreme the number of very poor people, according to frequent updates by the World Bank, declines very little if at all (graph 14). A person watching the television news and seeing some item about Bill Gates or Donald Trump on the business report and two minutes later a report on drought in north-east Africa can only conclude that the world has become more unequal. And, of course, they are right, even without a systematic quantitative measure of inequality.

Such measures, however, exist and the growing coverage of the data means that they are increasingly used to measure the distribution of income in the world as a whole. There are two common ways in which this is done: first, by comparing the extremes (usually the richest and the poorest 10 or 20 per cent of the population) (graph 10), and second, by using a measure of overall equality for the whole population (usually summarized by the Gini coefficient, a statistical measure of how much general incomes deviate from the average).

Either of these methods can be applied in two ways, which could be called international and global. International distribution, distribution between countries, takes into account only income differences between countries and not within countries. This amounts to assuming that all people receive the average income of their country. On this method the richest 10 per cent of the world's population are identified as the 600 million people who live in the richest countries as measured by their national income per head. Global distribution, however, is distribution between individuals irrespective of the economic level of the country where they live. So the richest 10 per cent in a global calculation would not include the whole population of the USA, for instance, but would consist of the 600 million individuals with the highest incomes, many of them in the USA, Europe and Japan, but some of them in Brazil, or even India.

Whether we measure the ratio of extremes or overall inequality, we find a huge long-run increase in international inequality during the 180 years for which data exist. In spite of numerous contradictory movements of individual countries (graph 114) the relative gap between rich and poor countries has widened (graph 115). The 'ratio of extremes' method shows that this widening still goes on, since the economic growth of the poorest countries continues to lag behind that of the richest ones. The measure of overall inequality, however, shows that since 1980 international inequality has been declining, since a number of economies in the middle of the distribution have grown very fast, especially China which accounts for most of the observed improvement.

Global as opposed to international measures of distribution show much greater inequality because they take into account inequalities within as well as between countries. The international ratio of the top to bottom 10 per cent of the population in 1997 is 30 to 1 while the equivalent global ratio is 63 to 1; the global measure of overall inequality (the Gini coefficient), too, is much higher than the international. But long-term changes in global distribution are impossible to measure because not enough comparable data exist about changes in internal income distribution. Studies do suggest that over the last twenty years the rise in internal inequality in many countries (graph 116) is enough to offset any lessening in overall inequality due to the growth of countries like China.

We are still far from having enough data to answer all the questions about changes in world equality and inequality. But this book is possible because of the enormous amount of incomplete but suggestive data which does exist. If one day we do have enough data to calculate total global inequality over the last quarter century it is a fair bet that we shall find that it has increased. But whether we do or whether we do not is not the important question. That question is whether the multiple inequalities which we observe today, regardless of how they have changed in the recent or remote past, reveal unacceptable social injustices. I hope that this book will help the reader to think about and discuss that question in a more informed way. My own reading of the evidence is that it would not give Sancho Panza's grandmother much cause to change the opinion she formed four hundred years ago.

References

(1) Edward Tufte, *The Visual Display of Quantitative Information*, Cheshire, Conn.: Graphics Press, 1983; *Envisioning Information*, Cheshire, Conn.: Graphics Press, 1990; *Visual Explanations*, Cheshire, Conn.: Graphics Press, 1997.

(2) Rudolf H. Strahm, *Pourquoi sont-ils si pauvres? Faits et chiffres en 84 tableaux sur les mécanismes du développement*, Boudry, Switzerland: Editions de la Baconnière, 1986.

(3) Norman Geras, 'Minimum Utopia: Ten Theses', in Leo Panitch and Colin Leys, eds, *Socialist Register 2000: Necessary and Unnecessary Utopias*, London: Merlin Press, 1999.

(4) Andrew Glyn and David Miliband, eds, *Paying for Inequality: The Economic Cost of Social Injustice*, London: IPPR/Rivers Oran Press, 1994.

(5) Fernando Fajnzylber, 'Las economías neoindustriales en el sistema centro–perifería de los ochenta', in Pedro Talavera Déniz, ed., *La Crisis Económica*

en América Latina, Barcelona: Sendai, 1991.

(6) Bob Sutcliffe, 'Development after Ecology', in V. Bhaskar and Andrew Glyn, eds, *The North the South and the Environment: Ecological Constraints and the Global Economy*, London: Easthscan/United Nations University, 1995.

(7) Andrea Boltho and Gianni Toniolo, 'Assessment: The Twentieth Century – Achievements, Failures, Lessons', *Oxford Review of Economic Policy*, vol. 15, no. 4, Winter 1999.

(8) Branko Milanovic, 'True World Income Distribution, 1988 and 1993: First Calculation Based on Household Surveys Alone', Research paper, Washington, D.C.: World Bank, 1999.

The data, the graphs and the text

Economic and social statistics have a bad reputation, being proverbially regarded as the superlative of lies. Ultimately the great majority of them derive from governments, and they can be wrong for many reasons: because they have been inefficiently collected, because they contain large elements of guesswork, because they have been deliberately falsified and because the concepts are bad. If we are to know something in detail about the world outside our own range of vision, however, there is really no alternative to using them as a tool, though one to be used critically. I have tried to select the data from the most reliable possible sources and have rejected data which seems especially defective. One feature of official statistics that may give one more faith in them is how often they yield conclusions that are very uncomfortable for the governments that produced them.

Aside from the enormous increase in the sheer quantity of data available in recent years, there have been major developments in the concepts that are used, such as new indices of development like the Human Development Index (graphs 21, 122) and the disability adjustment of life expectancy figures (graphs 23, 39). Figures about national income and product have changed more than most. Increasingly they are available in terms of purchasing power parity (see Glossary). This produces many changes in the relative levels of countries from more conventional national income figures (graph 8). For the reasons explained in the Glossary, in this book I have almost exclusively used the new purchasing power parity figures.

Each graph is accompanied by a page of text organized as follows:

- ● An explanation of the graph and how to read it.
- ○ Comments or elaborations on the content of the graph.
- ■ The brief source of the data used to produce the graph. Detailed sources are found in the section on Sources of the Data at the end.

Cross references to related previous or later graphs are indicated by arrows: e.g. ›93 or ‹21.

Terms which appear in **bold** in the text are defined in the Glossary of frequently used technical terms.

In the text and the graphs I employ pairs of terms such as North/South, developed countries/developing countries and rich countries/poor countries almost interchangeably. All of them seem to me to have virtues and drawbacks. I have no personal preference for any of them and, where appropriate, I have usually stuck to the version used in the source of the data.

The graphs were produced using a variety of computer programmes including Corel Presentations 8, Quatro Pro 8, Excel 97 and Map Maker Pro (whose creator, Eric Dudley, is responsible for the equal area projection used in several of the maps and whose Internet address is http://www.mapmaker.com). They were all finalized in Adobe Illustrator 8. If readers detect serious errors in the data presented I would appreciate this being drawn to my attention.

Glossary of frequently used technical terms

DALY The disability adjusted life year is a measure of personal and community welfare and of the burden of disease increasingly used by health economists, epidemiologists and the World Health Organization. It is explicitly designed to replace more economic measures of welfare such as level of income. The DALY is simply a year of life weighted by the level of disability. One year of life of a person with no disability has its full value. Otherwise the value of a year of life is reduced according to the level of disability. So when people are sick part of a DALY is lost. When they die the number of DALYs which are lost is the number of years between the age of their death and natural human life expectancy (assumed to be 80 years for men and 80.2 for women). When deaths are avoided or disabilities reduced, DALYs are gained, although, rather confusingly, in much of the literature the term 'DALY' is used to mean 'lost DALY'. Different medical expenditures can be compared in relation to the number of DALYs gained per unit of expenditure. And the importance of different health problems can be compared by estimating the number of DALYs lost from different diseases or produced by different risk factors. The human race, according to the WHO, lost a total of 1.4 billion DALYs in 1998: this is the sum of the years lost by people who die prematurely (the vast majority) and the partial years lost through illness and disability.

High, middle and low income countries This way of dividing the countries of the world is the one which has long been used by the World Bank in relation to national income figures, compared by means of exchange rates. In 1998 the dividing lines were $760 national income per head between low and middle income, and $9,360 between middle and high income. In some places in the book I have taken North to be equivalent to high income and South to be equivalent to middle plus low income countries respectively.

National and domestic income and product The national income or product is the way economists measure the amount of economic activity (the value of the production of goods and services) in a national economy. National product measures the value of the goods and services produced. National income measures the income (profits and wages) derived from these goods and services. The two are essentially the same measured from different angles of vision. Domestic product means the value of goods and services produced in a particular geographic area (usually a country). National product adjusts this figure by subtracting the amount transferred to foreigners for the production of goods and services (e.g. repatriation of profits on foreign investment) and adding the amount transferred by national citizens for the production of goods and services (e.g. the remittances of migrant workers). So, for example, in Ireland domestic product is higher than national product because of the large amount of foreign investment; while in Nicaragua the national product is larger than the domestic product because of the importance of migrant workers' remittances.

The national income or product can be measured gross or net (subtracting the estimated depreciation of the capital stock). Since the gross figure is much easier to measure, it is what we normally encounter. The most common measure of the size of an economy is the GDP, the gross domestic product. This was designed as a measure of economic activity but has come to be used (especially when calculated per head) as a measure of economic welfare and the standard of living. (see also **purchasing power parity**)

North/South This is a common if geographically inaccurate way of talking of two groups of countries in the world. The rich, industrialized countries are the North (North America, Japan, Western Europe and Australasia). The South refers to Africa, Asia, Latin America, the Caribbean and the Pacific. When I use the term in this book it always includes Eastern Europe but this is not universal practice elsewhere.

OECD (countries) The member countries of the OECD are those normally considered developed, rich and industrialized with the additions (for complex and different political reasons) of Mexico, Czechoslovakia, South Korea and Turkey. Usually when this category is used in this book it refers to the richer member countries only. It is thus a pseudonym for North or high-income countries. I have used it when the source materials of the data used it.

Purchasing power parity (ppp) This is a way of making international comparisons of incomes. Traditionally incomes in one country have been compared to those in another by converting them both into a common currency (which may or may not be that of one of the countries concerned) by means of the going exchange rate. But international travellers will know full well that the exchange rate is seldom equivalent to relative purchasing power. They find their money will buy more or less than it will at home (in Switzerland, and most richer countries, it buys less and in Morocco, and most poorer countries, it buys more). Indices of these levels of purchasing power are now available from the International Comparisons Project and are increasingly used by economists and international organizations. They provide a much more valid comparison of real income levels than the almost meaningless traditional method and so I have used them almost exclusively. The estimates available, however, are a long way from being perfect. PPP figures are sometimes called 'international dollars'.

Acknowledgements

I reiterate my thanks to those who helped me with the original Spanish edition: Carmen Perez Babot, Efrén Areskurrinaga, Elisa Sarsanedas, Itzíar Hernandez, Iñaki Gandariasbeitia, Julián Carranza, Charo Martinez, Alfonso Dubois, Pablo Bustelo, Joan Carlin and Scott Christie.

I am equally grateful to those who have helped with this English edition: especially to Robert Molteno of Zed Books, both for his encouragement and for his meticulous critique of the graphs; to my colleagues Andrew Glyn, Arthur MacEwan and Patxi Zabalo for help in searching out data and for valued comments.

Production, work and income

I

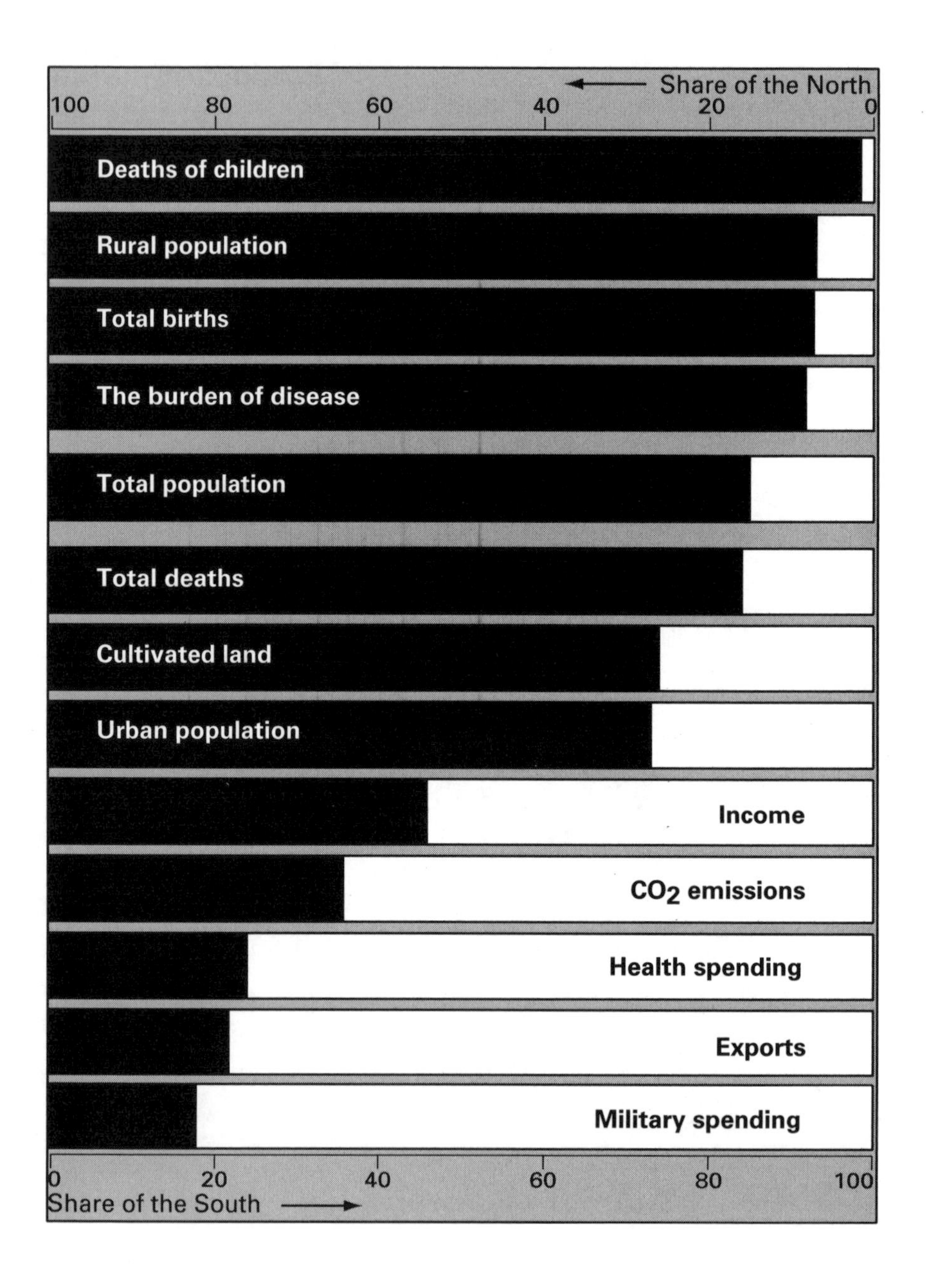

Share of the North
100
80
60
40
20
0
Deaths of children
Rural population
Total births
The burden of disease
Total population
Total deaths
Cultivated land
Urban population
Income
CO_2 emissions
Health spending
Exports
Military spending
0
20
40
60
80
100
Share of the South

Fundamental inequalities between North and South

● Each bar shows the distribution in about 1998 of a demographic or economic indicator between the countries of the North on the one hand (in white) and those of the South on the other (black) (see **North/South**).

○ The binary division of the world into North and South is very common. While too simplified for most purposes it nonetheless yields some important pointers to the nature of world inequalities. Start with the fifth bar, showing the division of the population between North and South. This is a simple yardstick for a 'fair share'. Equality between North and South would mean that the other bars were divided in the same proportions as population.

But they are not. From the first four bars we see that the South has more than its fair share of children's deaths, rural dwellers, total births and the overall burden of disease. But it has less than its fair share of total deaths, cultivated land, urban dwellers, income, CO_2 emissions, health spending, exports and military spending.

So we see here a world in which four-fifths of the economic resources devoted to health are spent in countries which have only 8 per cent of the burden of disease (› 31–41), in which the countries inhabited by only 20 per cent of the population receive 60 per cent of the world's income and are responsible for 80 per cent of world trade and an even higher proportion of military spending (›10–21, 70–77, 108–9).

Later we will see that, since these two groupings of countries are very heterogeneous, other more detailed ways of looking at inequality will show that the world is even more, indeed much more, unequal than it appears in this first simple approximation.

■ World Bank 1999a; WHO 1999; SIPRI 2000, UNCTAD 1999a.

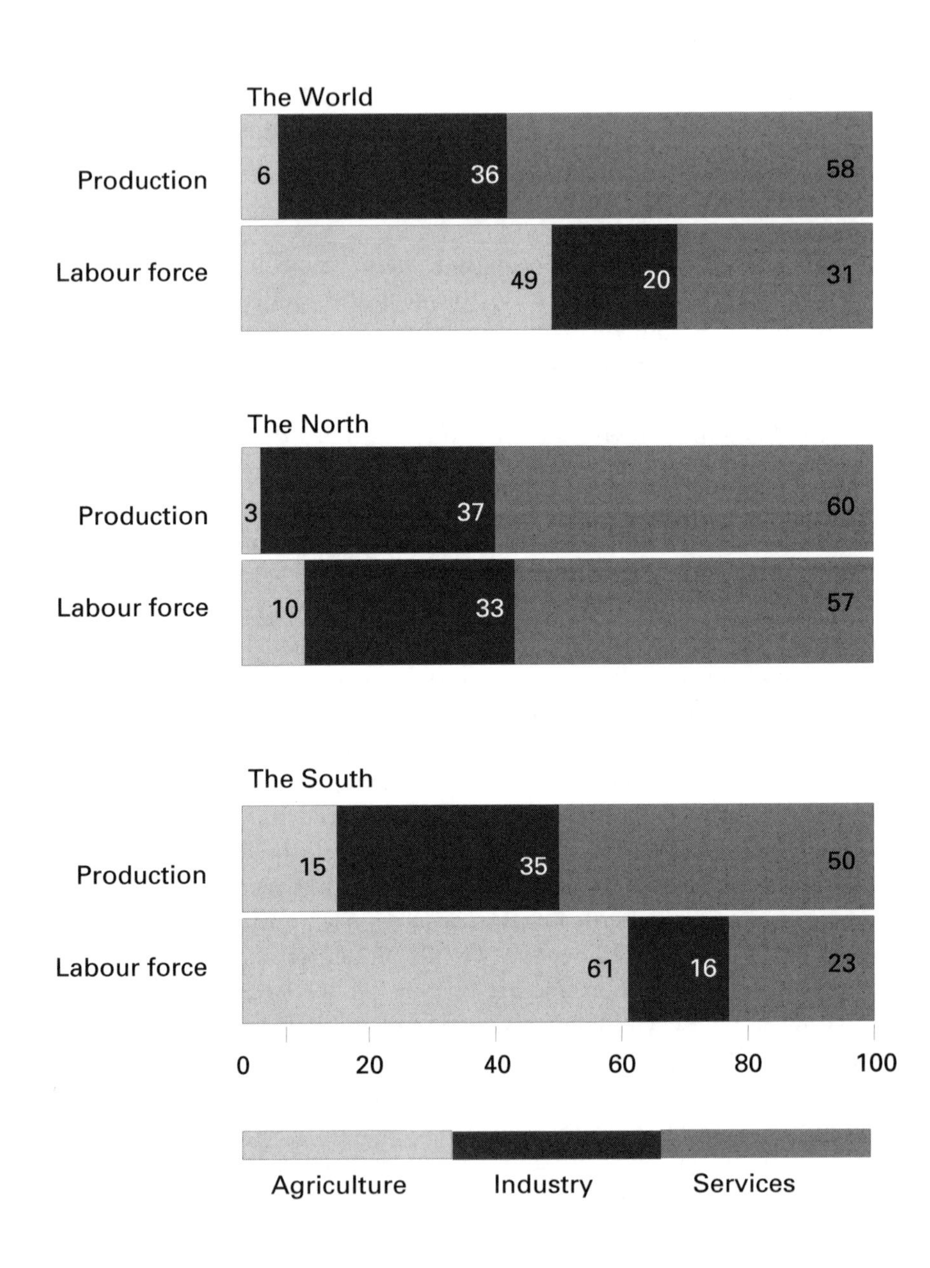

The World
Production
6
36
58
Labour force
49
20
31
The North
Production
3
37
60
Labour force
10
33
57
The South
Production
15
35
50
Labour force
61
16
23
0
20
40
60
80
100
Agriculture
Industry
Services

Contrasting structures of production and labour

● The bars divide the value of production and the size of the labour force between the three main economic sectors – agriculture, industry and services – for the world, the North and the South (see **North/South**).

○ The traditional way of viewing development attaches great importance to the way output and labour is divided between these three areas of activity. Development has been almost always defined as the structural transition of economies from a traditional state in which economic activity consists largely of low productivity agricultural production, via industrialization, to a state in which high-productivity industrial production comes to predominate over agriculture until finally it is far overtaken by services (a process sometimes called tertiarization). Some economists have argued that this process will begin by increasing inequality but will end up by reducing it. As we shall see, a lot of evidence contradicts this idea.

From this bar chart we see that there is a major difference between the economic structures of North and South. Proportionately six times more people work in agriculture in the South than in the North. And in the South proportionately only half as many people work in industry and services. Almost everywhere the percentage of the workforce in agriculture is higher than the percentage of output or income it produces. So, in this economic sense, labour is less productive in agriculture than in industry and services. In the South this difference in relative productivity is still a good deal greater than in the North.

■ World Bank 1999a.

Total world economic production, 1998 = US$30 trillion

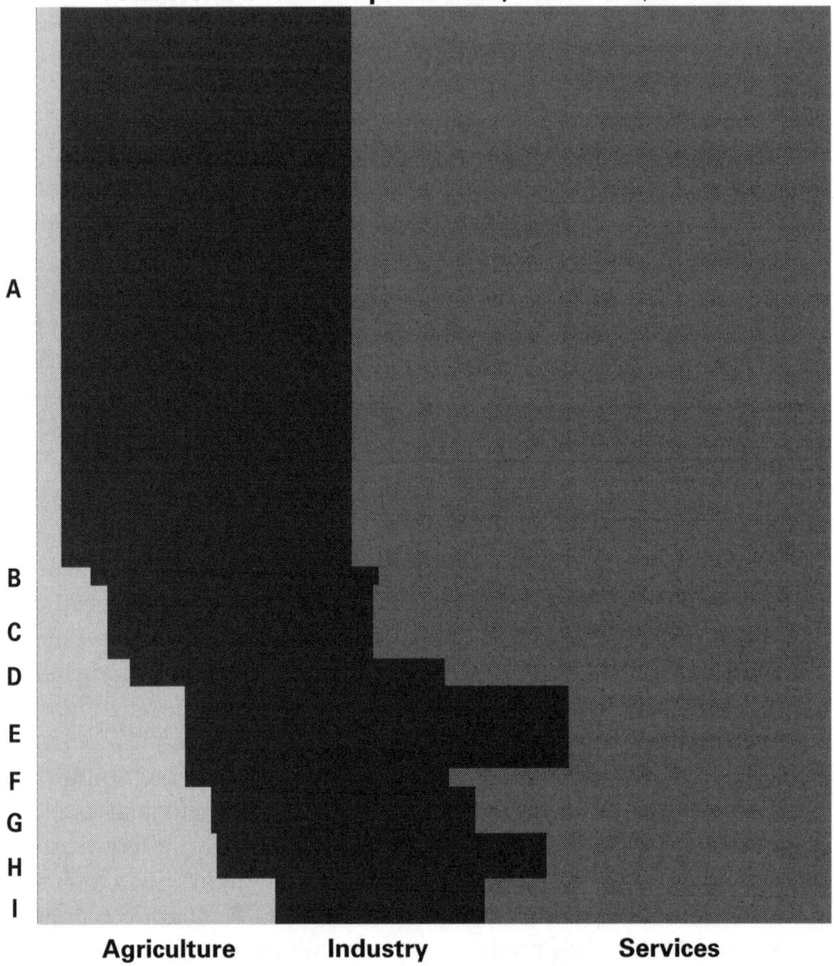

A

B

C

D

E

F

G

H

I

Agriculture **Industry** **Services**

A - Industrialized countries
B - East Asia (except China)
C - Latin America and Caribbean
D - Arab Countries
E - China
F - Sub-Saharan Africa
G - South–East Asia
H - Eastern Europe and ex-USSR
I - South Asia

● This chart shows the same information about the structure of production as the previous one but in more detail. The total area of the shaded rectangle represents the total value of world economic production in 1988 (about $30 trillion). The thickness of each layer of the rectangle (easily visible from the steps in the industry band) shows the weight of the corresponding area in world economic output and the shading shows the division into sectors. The areas are ordered in increasing order of importance of agriculture.

○ This graph gives a good idea both of the division of output between areas of the world and of the relative size of each sector. A number of things stand out:

- the overwhelming economic predominance of the industrial countries;
- the absence of a strong tendency for the share of industry to grow as that of agriculture falls;
- the especially large weight of the industrial sector in China and Eastern Europe and the ex-USSR, the legacy of communist economic policies;
- the clear tendency of services to grow as agriculture declines.

As typical locations of production of the world's output, the field and the factory are yielding to the office, the studio and the parlour.

■ World Bank 1999a.

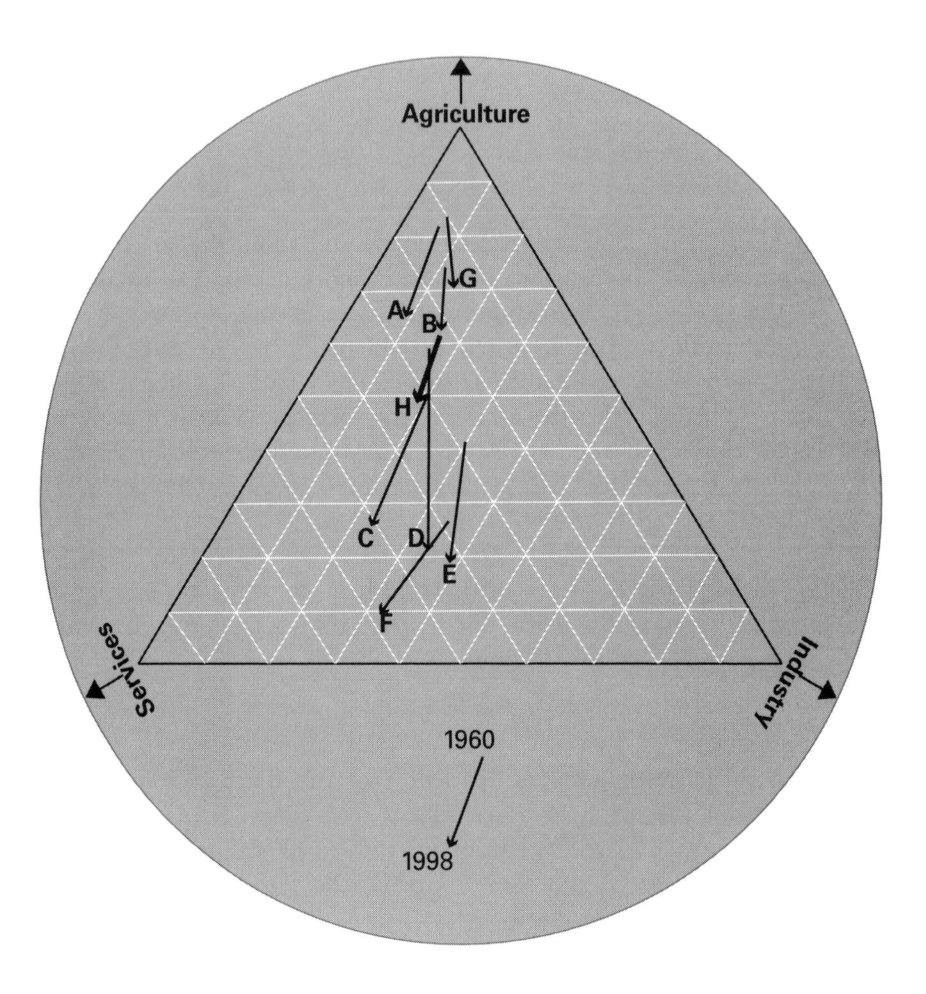

A - Sub-Saharan Africa
B - South Asia
C - Latin America
D - East Asia (except China)
E - Eastern Europe and ex-USSR
F - Industrialized Countries
G - China
H - World

Changes in the labour force, 1960–1998

● This triangular chart is another way of showing the percentages of the labour force working in industry, agriculture and services. Movements towards (or away from) the corners show increases (or decreases) in the percentage of the labour force in the indicated sector. At all points in the triangle the sum of the three shares is 100. The labelled lines show how the structure of the labour force has changed in the world and in seven of its main countries and regions between 1960 and 1998.

○ The sectoral division of labour in the world as a whole (line H) shows a clear tendency of movement during the last forty years: strongly away from agriculture, slightly towards industry and strongly towards services (in the triangle the line falls and at the same time inclines increasingly towards the left). The countries and regions in general follow this pattern but with differences and variations:

- services have greatest relative importance in the industrial countries;
- industry has greatest relative importance in Eastern Europe and the ex-USSR;
- Latin America (C) and East Asia (D) started in 1960 at about the same point on this graph and have since strongly diverged, East Asia having shown much more growth of industrial labour;
- China and Sub-Saharan Africa were also in a similar place forty years ago and have since diverged, with China industrializing more rapidly.

■ World Bank 1999a.

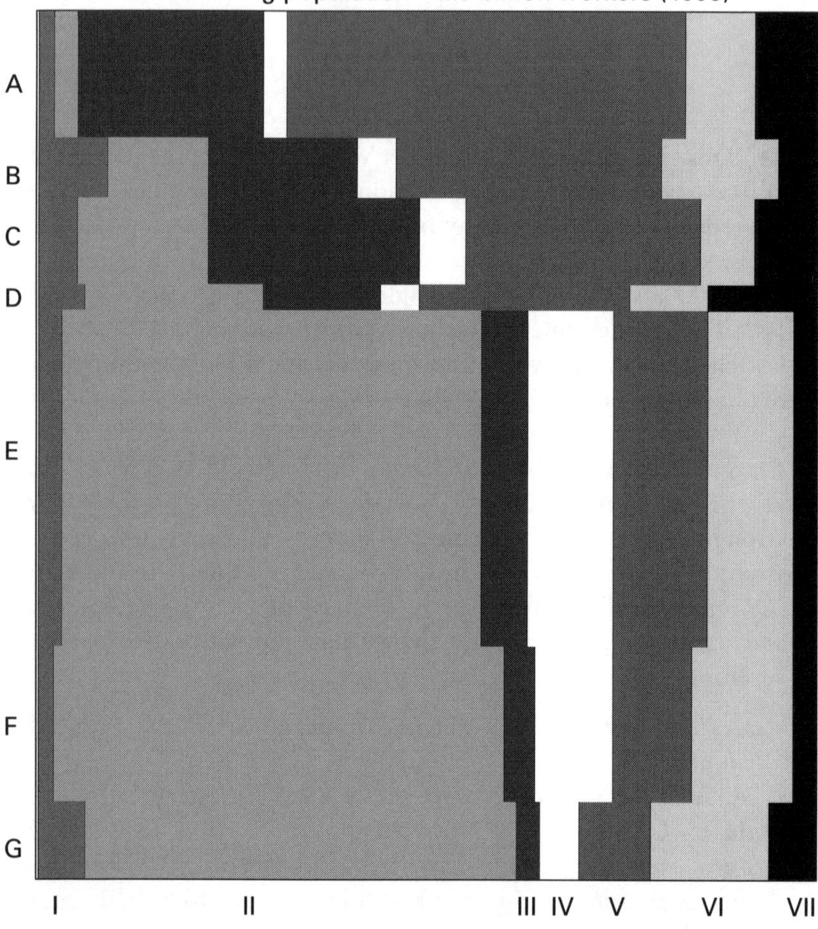

A. OECD
B. Latin America and Caribbean
C. Eastern Europe and Central Asia
D. West Asia and North Africa
E. East Asia and Pacific
F. South Asia
G. Sub-Saharan Africa

I. Agriculture (waged)
II. Agriculture (unwaged)
III. Industry (waged)
IV. Industry (unwaged)
V. Services (waged)
VI. Services (unwaged)
VII. Unemployment

Activities of the world's workers: field, factory and office

● Just as graph 3 showed the structure of production, this one shows the structure of the world's non-domestic labour force in 1995. The whole shaded area represents the 2,474 million people who were calculated to participate in the non-domestic labour force. The horizontal bands A–G are again scaled according to the relative numbers of workers; the vertical bands I–VII show the sectors in which people work (both with and without pay), and the part of the labour force which is unemployed.

○ As we saw in graph 1, the structure of the labour force reflects both the structure of output and the relative productivity of labour in different places and occupations. The addition of information about whether workers are waged or not provides a more detailed and revealing picture of the nature of human economic activity. The largest block of the human labour force by far is still people who work without wages in agriculture in the poorest groups of countries, in other words peasant farmers of various kinds.

The second largest is those who work for wages in service industries in the richer countries. And the third largest is waged industrial workers in the richer countries. In the richer countries waged labour predominates over unwaged in all three sectors. In the poorer ones unwaged dominates over waged labour. The graph therefore shows not only occupational differences but socio-economic ones as well.

■ Filmer 1995.

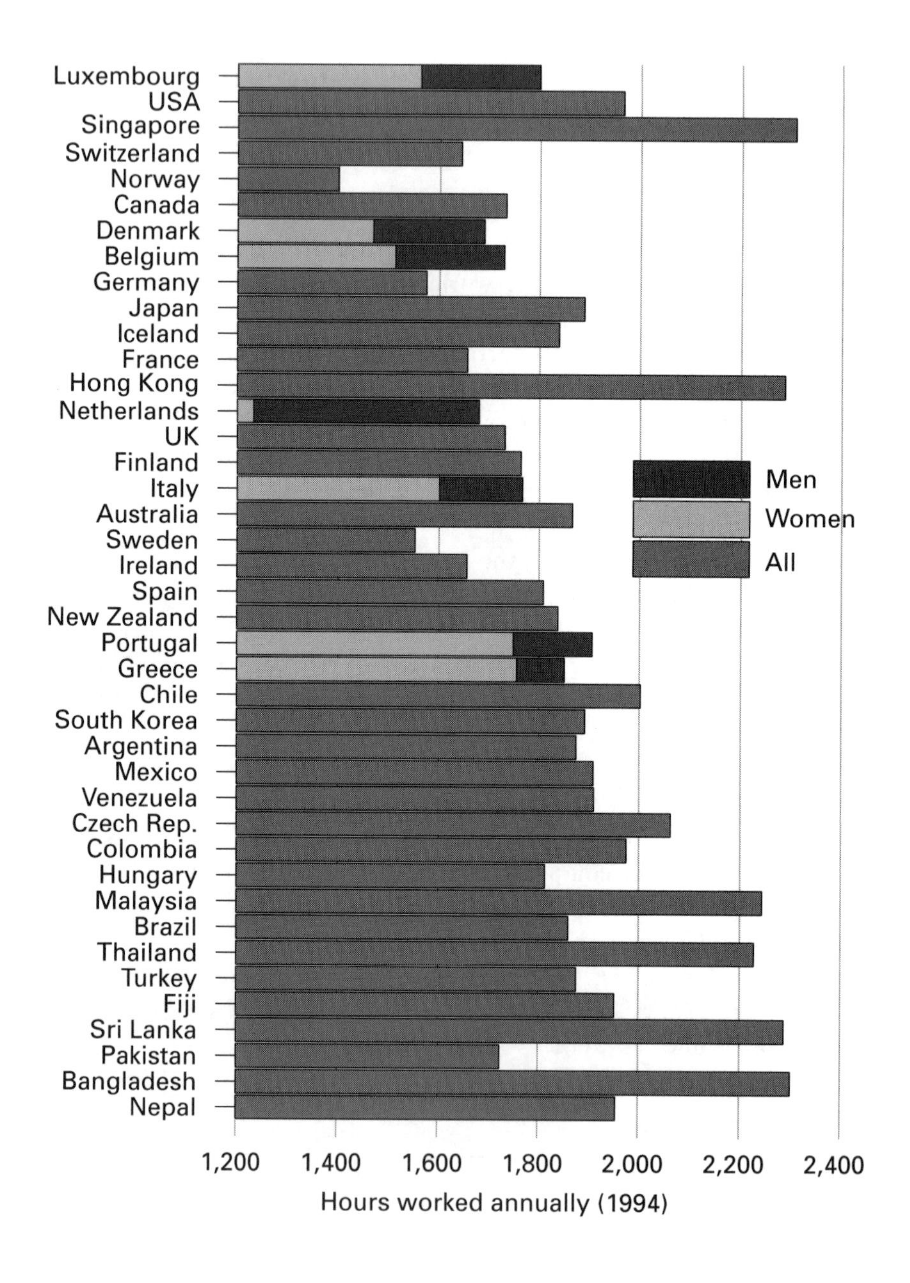

Luxembourg
USA
Singapore
Switzerland
Norway
Canada
Denmark
Belgium
Germany
Japan
Iceland
France
Hong Kong
Netherlands
UK
Finland
Italy
Australia
Sweden
Ireland
Spain
New Zealand
Portugal
Greece
Chile
South Korea
Argentina
Mexico
Venezuela
Czech Rep.
Colombia
Hungary
Malaysia
Brazil
Thailand
Turkey
Fiji
Sri Lanka
Pakistan
Bangladesh
Nepal
Men
Women
All
1,200
1,400
1,600
1,800
2,000
2,200
2,400
Hours worked annually (1994)

● These data come from a recent comparative study carried out by the International Labour Organization. The bars show the number of hours worked annually per employed person in forty-one developed and developing countries. They are shown in descending order of national income per head. For some countries there are different indicators for the hours of work of women and men workers, as indicated.

○ The quality of life as a whole is probably affected more than anything else by the nature of life at work. But indicators of the quantity and quality of working life are in general a well-kept secret. These data refer not to all work but to average hours actually worked in paid jobs outside the home. Countries are listed in descending order of national income per head and this seems to show a slight negative relationship between the wealth of a country and the number of hours which people work, suggesting in general that greater wealth goes with greater work productivity and more leisure. Yet there are clear exceptions. For instance, working hours in the USA are considerably longer than in most other developed countries. And the countries where working hours are longest are Hong Kong and Singapore. This sheds some new light on countries which are often regarded as models. The economic superiority of the USA over Western Europe would be lower if measured in productivity per hour of work than in level of income. And the economic 'miracles' of some developing economies may consist as much in making their citizens work miraculously long hours as in miraculously transforming productivity. Working hours are shortest in the Scandinavian countries, closely followed by others in Western Europe.

■ ILO 1999.

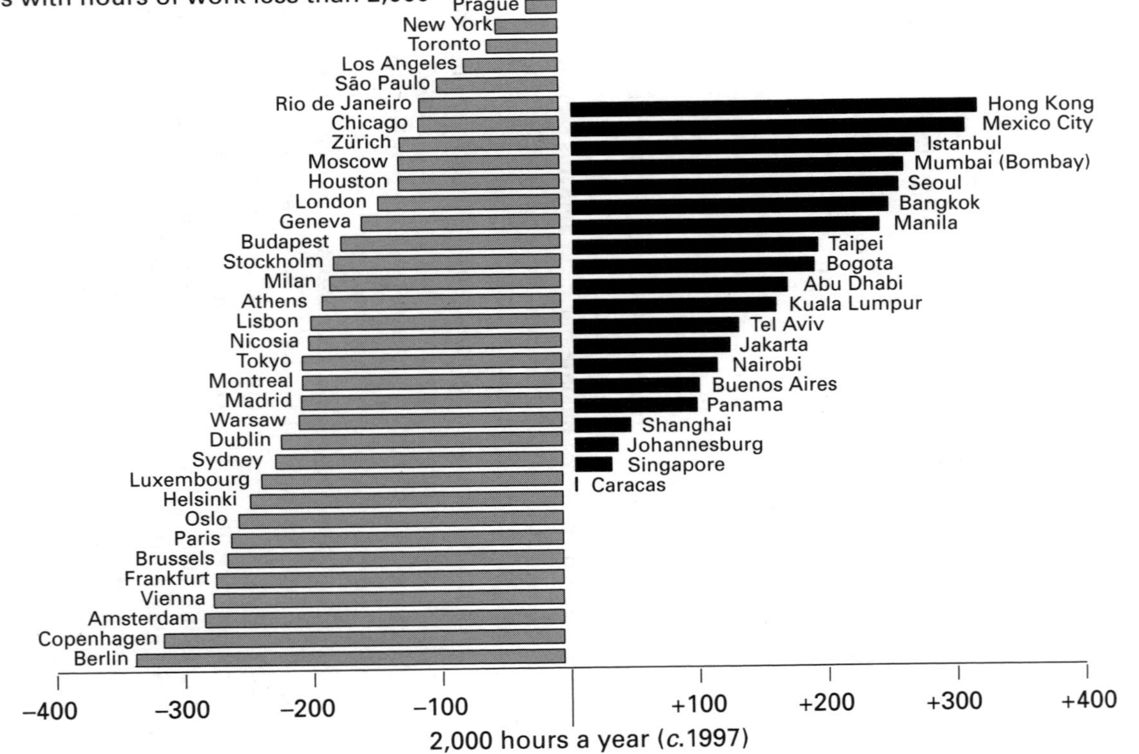
Cities with hours of work less than 2,000
Cities with hours of work more than 2,000
Bahrain
Prague
New York
Toronto
Los Angeles
São Paulo
Rio de Janeiro
Chicago
Zürich
Moscow
Houston
London
Geneva
Budapest
Stockholm
Milan
Athens
Lisbon
Nicosia
Tokyo
Montreal
Madrid
Warsaw
Dublin
Sydney
Luxembourg
Helsinki
Oslo
Paris
Brussels
Frankfurt
Vienna
Amsterdam
Copenhagen
Berlin
Hong Kong
Mexico City
Istanbul
Mumbai (Bombay)
Seoul
Bangkok
Manila
Taipei
Bogota
Abu Dhabi
Kuala Lumpur
Tel Aviv
Jakarta
Nairobi
Buenos Aires
Panama
Shanghai
Johannesburg
Singapore
Caracas
–400
–300
–200
–100
+100
+200
+300
+400
2,000 hours a year (c.1997)

● These data come from a study of workers in cities carried out by the research department of the Union des Banques Suisses. The graph divides cities according to the extent to which hours were found to be above or below 2,000 a year. So in Hong Kong people work about 2,300 and in Berlin about 1,660. The data refer mainly to employees in manufacturing industry.

○ The data in this graph are from a different and less systematic study than the previous one, but their results are in many ways consistent with it. They also show higher working hours in North America than in Europe, and even than in Latin America. They confirm that the longest working hours are to be found in Asia and that they are particularly long in some of the countries which have been considered economic miracles. There is one difference here, however, in that this study, unlike the ILO one (‹6), finds exceptionally long working hours in Seoul, the capital of South Korea, usually regarded as one of the most economically successful developing countries, which has recently been elevated to membership of the 'rich countries' club', the **OECD**. The ILO findings are also different in the case of Turkey.

The most striking message of this graph is the very wide range of working hours that exist in the world. An average worker in Hong Kong, Mexico City or Istanbul works about 600 more hours a year than her or his counterpart in Berlin or Copenhagen or Amsterdam. That difference is the result both of longer hours every day and of fewer days of holiday – factors which, most people would agree, make an enormous difference and should be accorded greater importance in measuring differences in the quality of life.

■ UBS 1997.

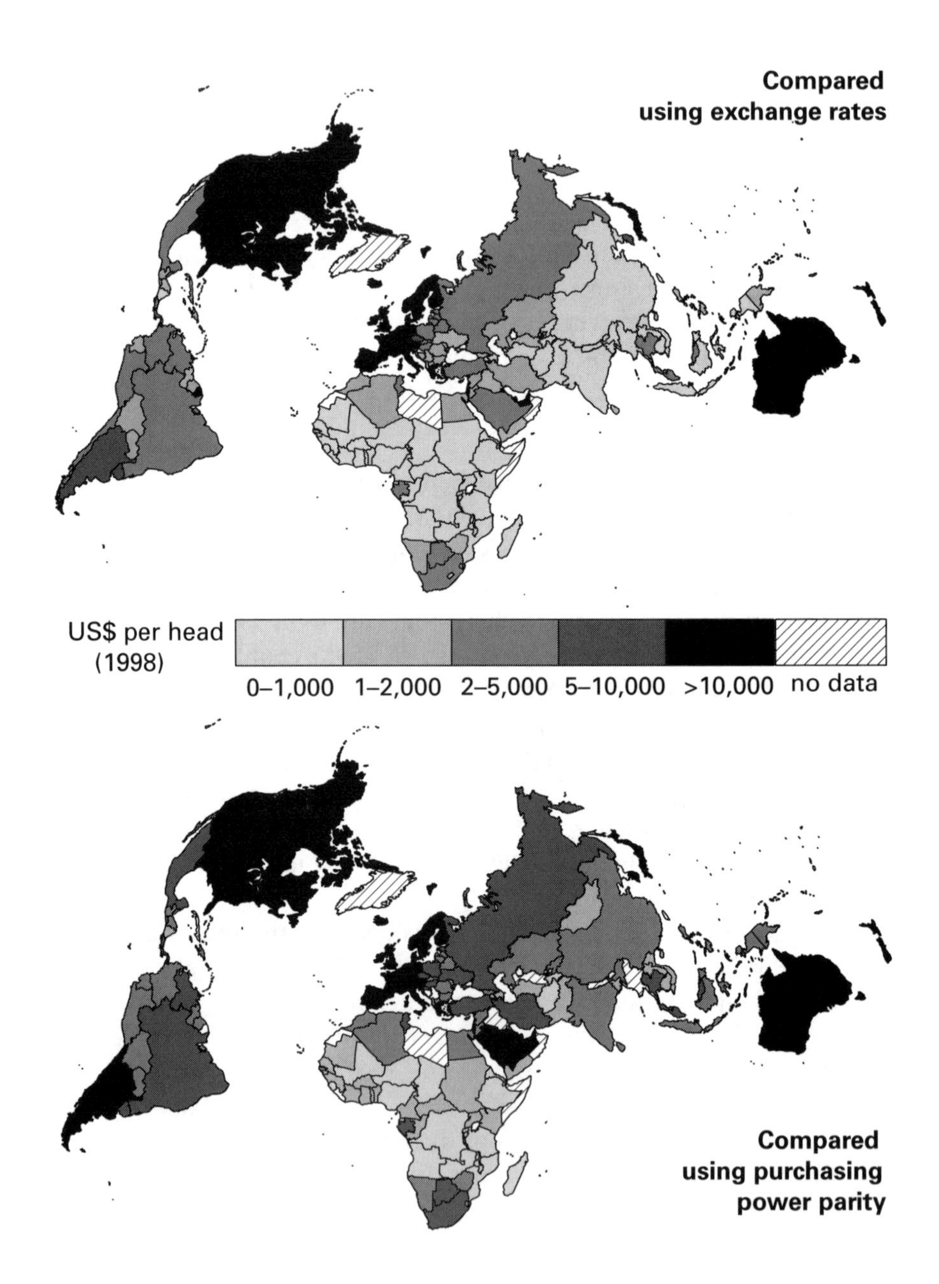
Compared
using exchange rates
US$ per head
(1998)
0–1,000
1–2,000
2–5,000
5–10,000
>10,000
no data
Compared
using purchasing
power parity

Two ways of comparing incomes

● These two maps show the level of individual countries' gross national product per head converted to US dollars using two different methods: by exchange rates, and by an estimate of **purchasing power parity**.

○ These two maps show, among other things, the hazards of the economic statistics that we encounter every day. In the last few years estimates of the national income of countries compared by the method of **purchasing power parity** have become generally available. So they now circulate side by side with estimates made using exchange rates, causing immense confusion. Statisticians, economists, international organizations, the press and politicians now may quote figures without explaining which method they are using. Yet the two methods give considerably different pictures of the economic nature of the world. To give an example, it is often said that Japan is the second largest economy in the world, especially when Western governments are trying to put pressure on Japan to 'assume its international responsibilities' and reflate its flagging economy. This is true if using exchange rate conversions. But if purchasing power parity conversions are used the world's second largest economy, by a very long way, is China and not Japan.

In this book, however, I have always used ppp figures when they are available because in principle they allow a more valid comparison. The ppp method raises the estimates of income levels of many poorer countries (especially the large Asian countries), as is shown by comparing these two maps. This way of measuring inequality is still unfamiliar, but it gives a much more meaningful picture of the degree of international income inequality and of changes in it over time.

■ World Bank 2000b.

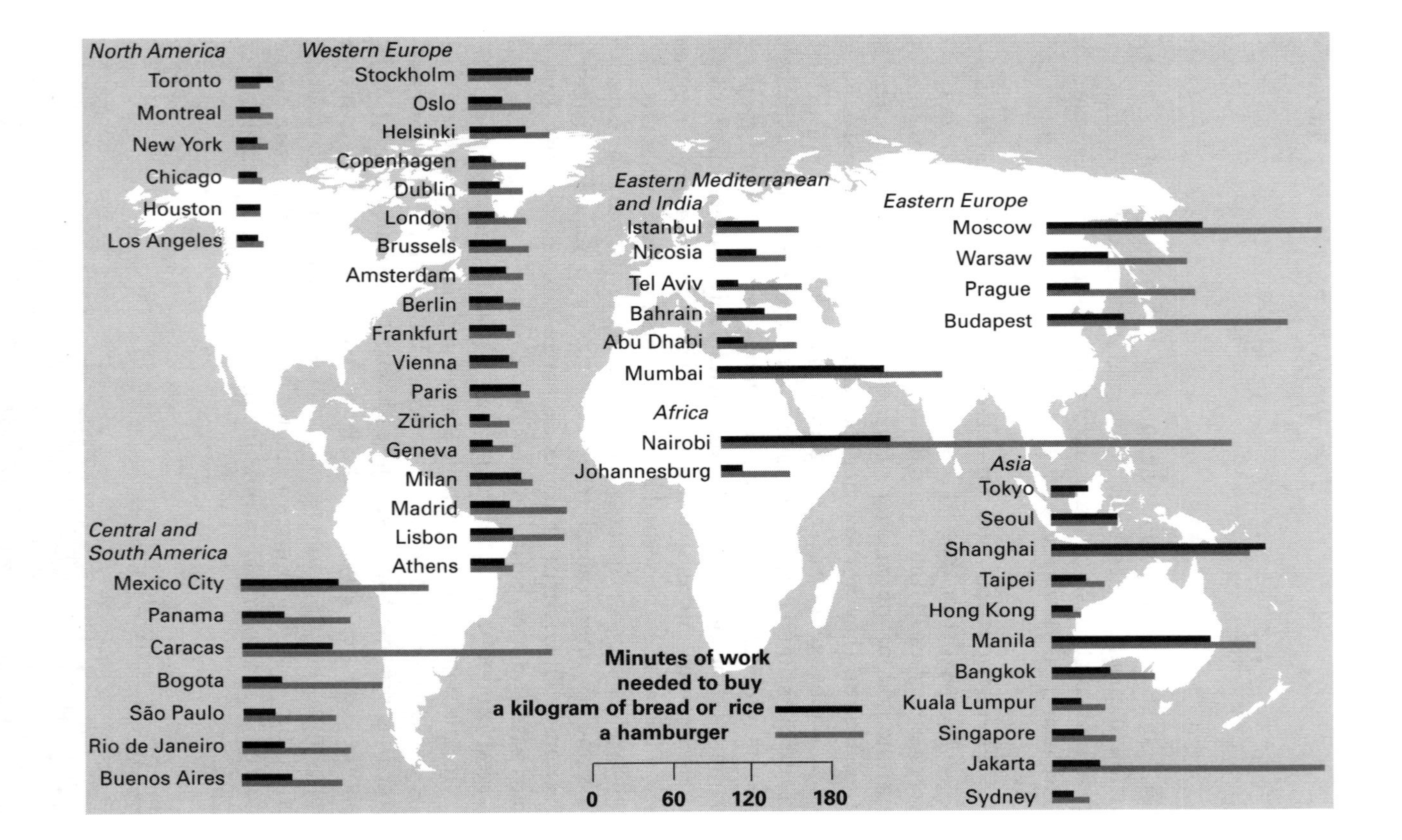
North America
Toronto
Montreal
New York
Chicago
Houston
Los Angeles
Western Europe
Stockholm
Oslo
Helsinki
Copenhagen
Dublin
London
Brussels
Amsterdam
Berlin
Frankfurt
Vienna
Paris
Zürich
Geneva
Milan
Madrid
Lisbon
Athens
Central and South America
Mexico City
Panama
Caracas
Bogota
São Paulo
Rio de Janeiro
Buenos Aires
Eastern Mediterranean and India
Istanbul
Nicosia
Tel Aviv
Bahrain
Abu Dhabi
Mumbai
Africa
Nairobi
Johannesburg
Eastern Europe
Moscow
Warsaw
Prague
Budapest
Asia
Tokyo
Seoul
Shanghai
Taipei
Hong Kong
Manila
Bangkok
Kuala Lumpur
Singapore
Jakarta
Sydney
Minutes of work needed to buy
a kilogram of bread or rice
a hamburger
0
60
120
180

What work will buy, 1998

● This chart also tries to measure real differences of purchasing power between people in different countries, but much less systematically than the figures on which the lower map in graph 8 is based. This study investigated how much working time at the average wage is needed in different cities to buy a kilo of a basic foodstuff (bread or rice) and a food luxury (a hamburger). The amount of time is shown by the length of the black and grey bars respectively. The study is flawed by using only two products, which may not have the same significance in all places.

○ For those baffled by economic measures of international differences in welfare this chart may be instinctively easier to understand. It is a simplified way of seeing the concept of differences in the material standard of living. It helps to demystify the world of prices, income, pounds and dollars. The fact is that consumption requires work. And our concept of luxuries and necessities can be converted into the universal currency of work. Something is a luxury because we need to work a very long time to earn enough to buy it. The short bars in this chart for the cities in developed countries means that very little time needs to be worked to buy a kilo of rice or bread, or even a less basic food item like a hamburger; none of them is a luxury. The very few cities in poor countries which are included in this study show that the bars are longer; in other words a person must work longer to get food. It appears that the difference in the length of the bars between rich and poor countries may be relatively greater in the case of bread or rice than in the case of hamburgers. This means not that the hamburger is less of a luxury but that even basic items are, in poor economic circumstances, luxuries.

■ UBS 1997.

The inequality of income II

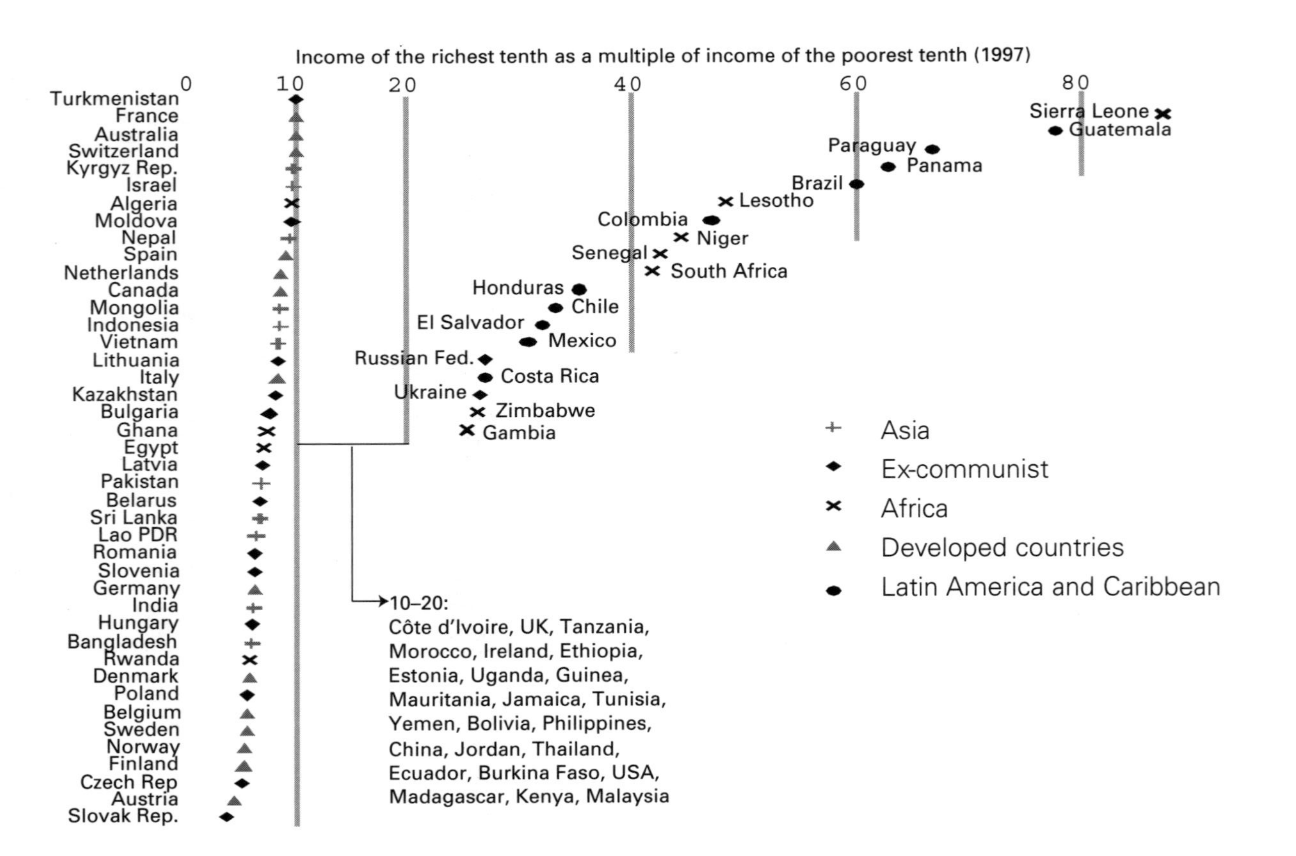
Income of the richest tenth as a multiple of income of the poorest tenth (1997)
0
10
20
40
60
80
Turkmenistan
France
Australia
Switzerland
Kyrgyz Rep.
Israel
Algeria
Moldova
Nepal
Spain
Netherlands
Canada
Mongolia
Indonesia
Vietnam
Lithuania
Italy
Kazakhstan
Bulgaria
Ghana
Egypt
Latvia
Pakistan
Belarus
Sri Lanka
Lao PDR
Romania
Slovenia
Germany
India
Hungary
Bangladesh
Rwanda
Denmark
Poland
Belgium
Sweden
Norway
Finland
Czech Rep
Austria
Slovak Rep.
10–20:
Côte d'Ivoire, UK, Tanzania,
Morocco, Ireland, Ethiopia,
Estonia, Uganda, Guinea,
Mauritania, Jamaica, Tunisia,
Yemen, Bolivia, Philippines,
China, Jordan, Thailand,
Ecuador, Burkina Faso, USA,
Madagascar, Kenya, Malaysia
Russian Fed.
Costa Rica
Ukraine
Zimbabwe
Gambia
Honduras
Chile
El Salvador
Mexico
Senegal
South Africa
Colombia
Niger
Lesotho
Brazil
Paraguay
Panama
Sierra Leone
Guatemala
Asia
Ex-communist
Africa
Developed countries
Latin America and Caribbean

Contrasting degrees of income inequality

● This diagram shows information about the comparative degree of inequality of income within countries. The points represent the ratio of the income (after taxes and benefits) of the richest tenth to the poorest tenth of national populations in the mid-1990s. For reasons of space the countries in the intermediate group (with rich to poor ratios of 10–20) are simply listed in ascending order without the value of the ratio being recorded.

○ Comparing the richest to the poorest groups of a society is a comprehensible way of measuring its inequality. At the extreme, the figures show that in Sierra Leone the richest tenth receive over 80 times the income of the poorest tenth. In a number of Latin American and African countries the ratio exceeds 40. At the other extreme are a few countries with a ratio between rich and poor of less than 5. Most of those with less than 10 are countries in Central and Western Europe and in East and South Asia, although a number of these have recently become more unequal (›116).

It is common to read disparaging references in the Western press to the inequality in a country such as India, so it is salutary to note that, by this useful measure, inequality in the UK and in the USA is much greater than in India. In both the UK and the USA this measure of inequality has risen in recent decades and they are both now in the intermediate group, having previously been in the lower group.

For the world as a whole the ratio between the richest and poorest tenths of the population is 63 (see p. 8). In other words the world by this measure is more unequal than all except two of its component countries.

■ World Bank 2000b.

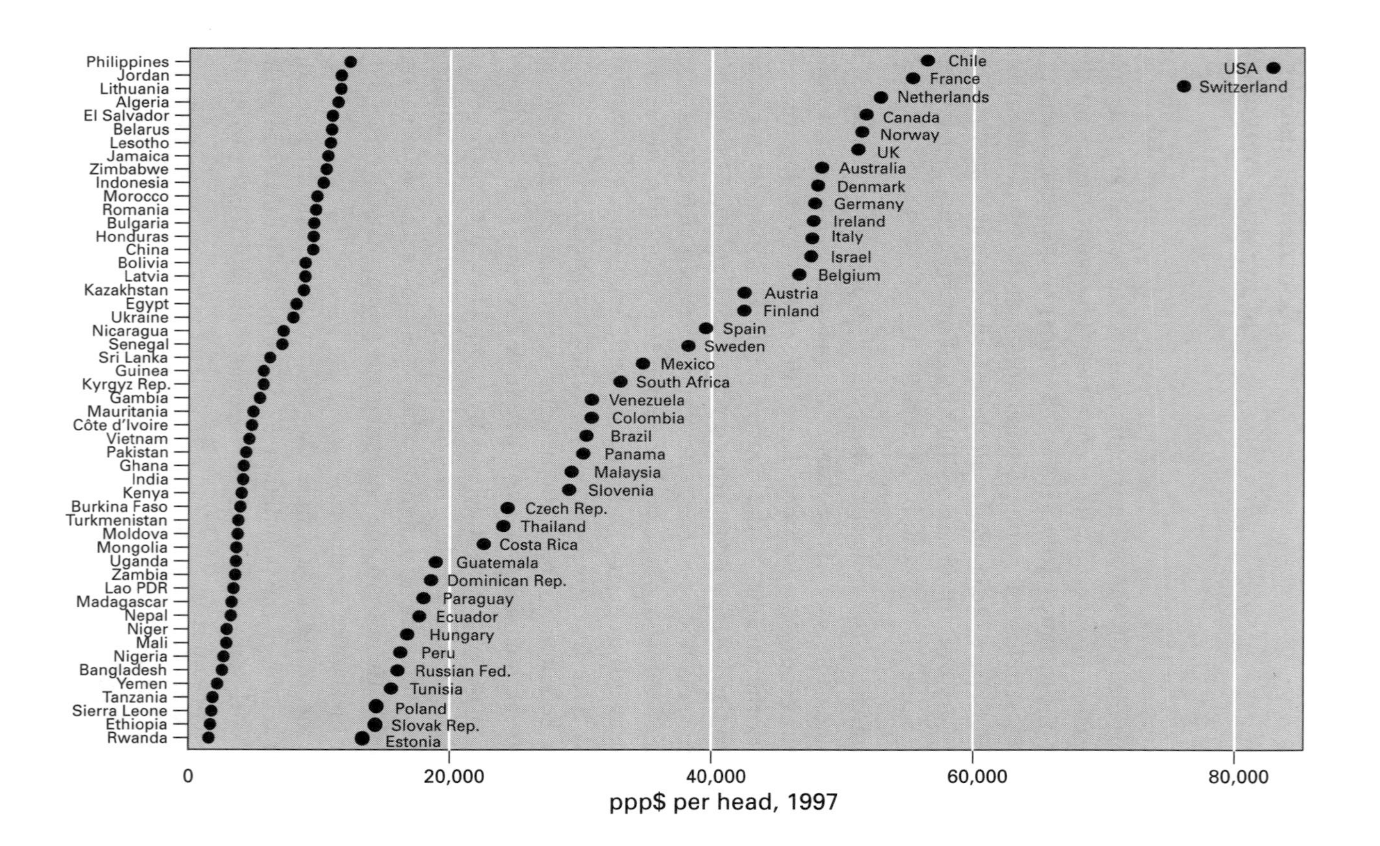
Philippines
Jordan
Lithuania
Algeria
El Salvador
Belarus
Lesotho
Jamaica
Zimbabwe
Indonesia
Morocco
Romania
Bulgaria
Honduras
China
Bolivia
Latvia
Kazakhstan
Egypt
Ukraine
Nicaragua
Senegal
Sri Lanka
Guinea
Kyrgyz Rep.
Gambia
Mauritania
Côte d'Ivoire
Vietnam
Pakistan
Ghana
India
Kenya
Burkina Faso
Turkmenistan
Moldova
Mongolia
Uganda
Zambia
Lao PDR
Madagascar
Nepal
Niger
Mali
Nigeria
Bangladesh
Yemen
Tanzania
Sierra Leone
Ethiopia
Rwanda
Chile
France
Netherlands
Canada
Norway
UK
Australia
Denmark
Germany
Ireland
Italy
Israel
Belgium
Austria
Finland
Spain
Sweden
Mexico
South Africa
Venezuela
Colombia
Brazil
Panama
Malaysia
Slovenia
Czech Rep.
Thailand
Costa Rica
Guatemala
Dominican Rep.
Paraguay
Ecuador
Hungary
Peru
Russian Fed.
Tunisia
Poland
Slovak Rep.
Estonia
USA
Switzerland
0
20,000
40,000
60,000
80,000
ppp$ per head, 1997

● Combining the **PPP** estimates of **national income** per head and the World Bank's estimates of its distribution within countries we can impute the average level of income of the richest 10 per cent of national populations, indicated by a circle on the graph.

○ Here we see a comparison between the incomes of the rich of each country. The approximately 27 million richest US citizens receive over $80,000 a year per person. The figures for the USA and Switzerland are far higher than those of other countries, with Chile a surprising third in this league. At the low end of this scale an average member of the richest tenth of the population in most African countries receives less than an average poor person in the most developed countries.

The order of countries in this chart is different from their order in terms of the average income for the whole of the population. For instance the richest tenth of the population of Chile is richer than the richest tenth of the population of the UK, although the UK has a considerably higher level of income per head overall than Chile. The income of the richest Swedes is very similar to that of the richest Mexicans, even though Sweden as a whole has a much higher level of income per head than Mexico. These curious facts draw attention to differences from country to country in the treatment of the rich, or in other words to differences in the internal distribution of income. This is affected by many aspects of the society and economy: the concentration of property ownership, the degree of social reverence for riches, conventions about the acceptable ranges of wage and salary differentials, the existence of financial institutions like stock markets and the degree of progressiveness in the tax system.

■ World Bank 2000b.

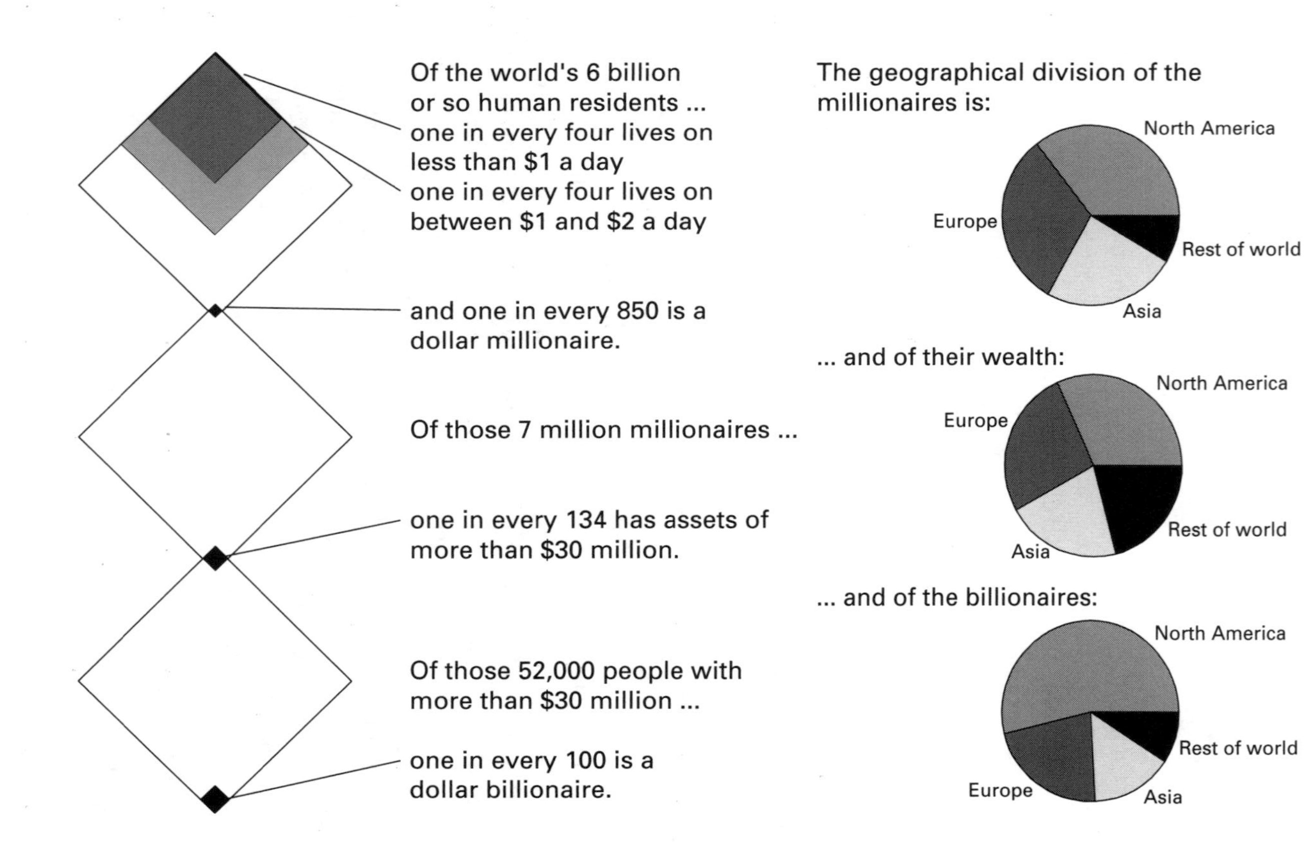
Of the world's 6 billion or so human residents ...
one in every four lives on less than $1 a day
one in every four lives on between $1 and $2 a day
and one in every 850 is a dollar millionaire.
Of those 7 million millionaires ...
one in every 134 has assets of more than $30 million.
Of those 52,000 people with more than $30 million ...
one in every 100 is a dollar billionaire.
The geographical division of the millionaires is:
North America
Europe
Rest of world
Asia
... and of their wealth:
North America
Europe
Rest of world
Asia
... and of the billionaires:
North America
Rest of world
Europe
Asia

● The top diamond represents the whole population of the world and shows the relative number of poor people (with less than $1 and $2 a day) and of millionaires (with wealth of more than $1 million). The second represents the population of millionaires and indicates the proportion of them who have more than $30 million; and the third shows billionaires as a share of those with more than $30 million. Rich is here defined in terms of wealth (possession of assets) not income.

○ Not much is known about the rich. This is because they find it politic to be secretive in order to protect their wealth and also because social science and politics have defined poverty as a pathological symptom of society but, illogically, not riches.

We owe the limited amount of information about the rich summarized here not to social science but to Merrill Lynch, a firm of financial advisors. It hopes its *World Wealth Report* will enable it to serve profitably those whom it calls 'High Net Worth Individuals' (HNWIs), a euphemism no doubt designed to expunge the memory of adjectives traditionally associated with the rich such as 'filthy' and 'stinking'.

Of the world's 7 million millionaires (1 in every 850 of the population) 2.5 million are estimated to be in North America (making them almost 1 per cent of the population there), 2.2 million in Europe, 1.7 million in Asia and about 650,000 in the rest of the world.

With millionaires so common it is not surprising that Merrill Lynch takes increasing interest in the even richer 52,000 U-HNWIs (Ultra-High Net Worth Individuals) and their elite, the estimated 512 billionaires.

■ Merrill Lynch and Gemini Consulting 2000.

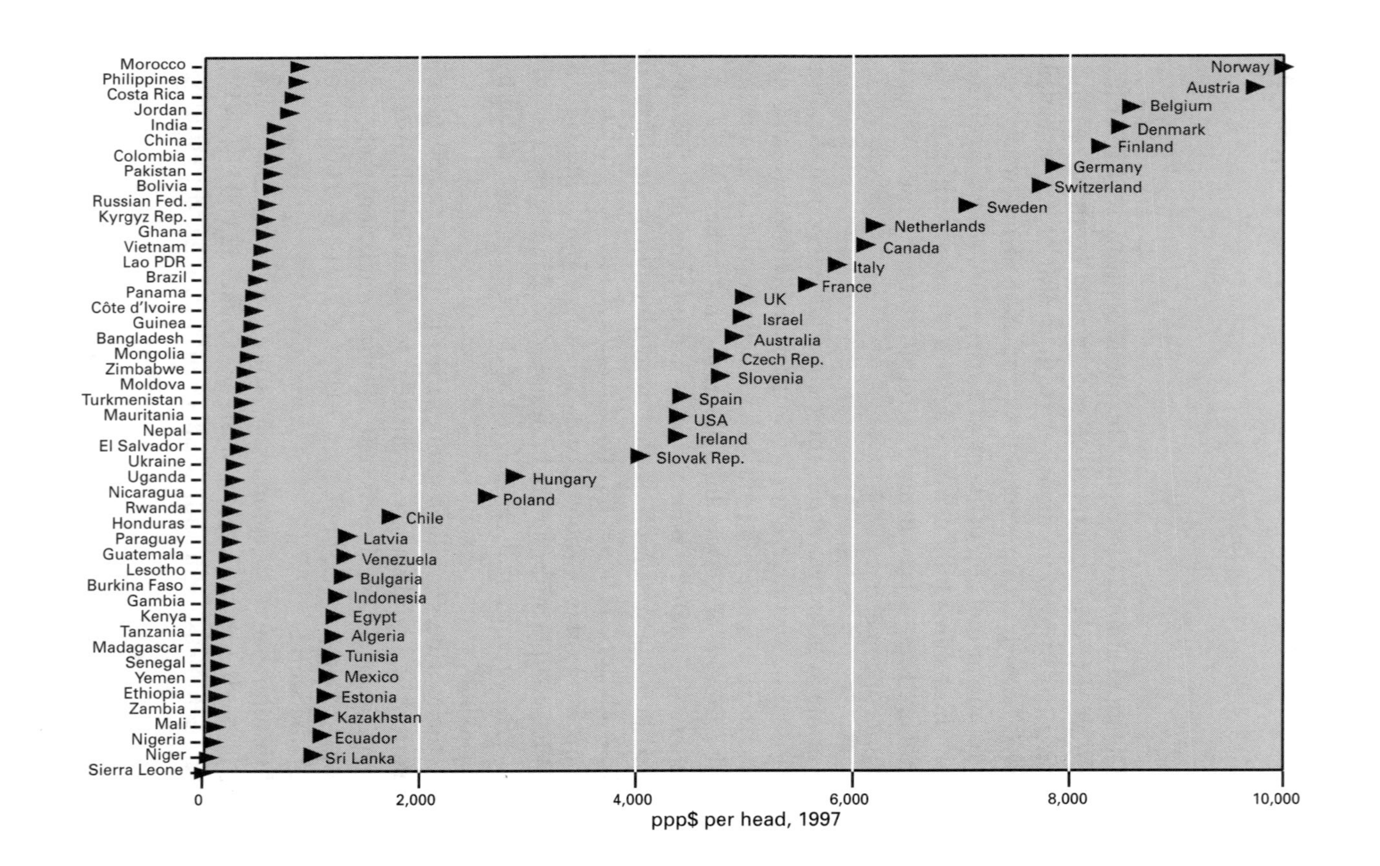
Morocco
Philippines
Costa Rica
Jordan
India
China
Colombia
Pakistan
Bolivia
Russian Fed.
Kyrgyz Rep.
Ghana
Vietnam
Lao PDR
Brazil
Panama
Côte d'Ivoire
Guinea
Bangladesh
Mongolia
Zimbabwe
Moldova
Turkmenistan
Mauritania
Nepal
El Salvador
Ukraine
Uganda
Nicaragua
Rwanda
Honduras
Paraguay
Guatemala
Lesotho
Burkina Faso
Gambia
Kenya
Tanzania
Madagascar
Senegal
Yemen
Ethiopia
Zambia
Mali
Nigeria
Niger
Sierra Leone
Norway
Austria
Belgium
Denmark
Finland
Germany
Switzerland
Sweden
Netherlands
Canada
Italy
France
UK
Israel
Australia
Czech Rep.
Slovenia
Spain
USA
Ireland
Slovak Rep.
Hungary
Poland
Chile
Latvia
Venezuela
Bulgaria
Indonesia
Egypt
Algeria
Tunisia
Mexico
Estonia
Kazakhstan
Ecuador
Sri Lanka
0
2,000
4,000
6,000
8,000
10,000
ppp$ per head, 1997

● Each triangle represents the average income of the poorest 10 per cent of the population, calculated in the same way as in graph 11 for the income of the richest 10 per cent.

○ The inequality between the position of the poor in different countries, clearly visible here, is much more important than the national differences between the rich. In all countries the rich live well. In many countries the poor cannot even survive physically. Here we find, not surprisingly, that in general there is coincidence between the order of countries in terms of average income and their order in terms of the income of the poor. But there are still striking differences that show how economic customs and policies can create differences between societies which in other respects seem economically, politically and ideologically very similar. The Scandinavian countries, along with Austria, Switzerland, Belgium and Germany are relatively rich countries with a relatively equal distribution of income. And so the average person even in the poorest 10 per cent of the population has an income which in global terms is high. In other rich but less egalitarian countries (which include the UK) the poorest part of the population fare worse. And it is striking that in the richest country of all, the USA, the poorest part of the population are poorer than in almost any other developed country. You are marginally better off if you are poor in Spain than if you are poor in the USA.

These figures in principle compare absolute levels of deprivation. But the social condition of poverty is not adequately described by the level of material consumption; it also comprises deprivation relative to those who share the same social and economic space.

■ World Bank 1999a.

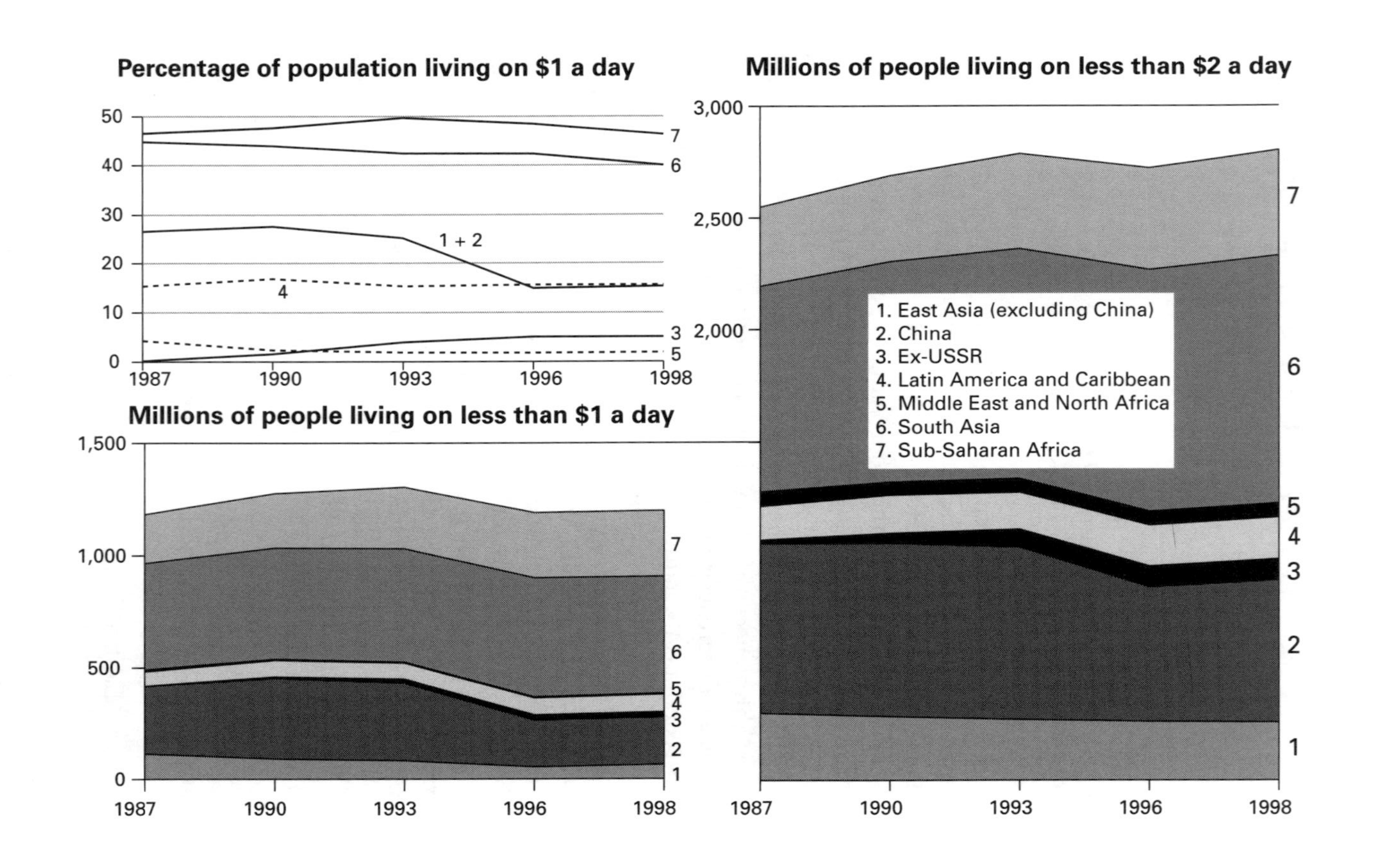
Percentage of population living on $1 a day
50
40
30
20
10
0
1987
1990
1993
1996
1998
7
6
1 + 2
4
3
5
Millions of people living on less than $1 a day
1,500
1,000
500
0
1987
1990
1993
1996
1998
7
6
5
4
3
2
1
Millions of people living on less than $2 a day
3,000
2,500
2,000
1987
1990
1993
1996
1998
1. East Asia (excluding China)
2. China
3. Ex-USSR
4. Latin America and Caribbean
5. Middle East and North Africa
6. South Asia
7. Sub-Saharan Africa
7
6
5
4
3
2
1

● The graphs are based on the World Bank's concept of income poverty, which estimates the numbers of people in the world living below a given level of income. The two charts on the left show the percentage of the population and the absolute numbers of people who are estimated to have incomes below $1 a day and how this has evolved between 1987 and 1998. The right-hand chart shows the absolute numbers of people living on less than $2 a day.

○ These figures, rapidly becoming the most quoted economic figures in the world, show that about one-quarter of the world's population is below the lower poverty line ($1 a day) and about half below the upper poverty line ($2). The percentages have declined very slowly in the two poorest regions, South Asia and sub-Saharan Africa, and quite sharply in China and other parts of East Asia; but they have risen sharply in the countries of the former Soviet Union. Over the ten years covered by these estimates the total number of poor people in the world, according to this absolute definition, has either stayed about the same or risen.

Although these figures are calculated in terms of **purchasing power parity** they mean very different things in different places and social conditions. In many places a person with no more than $2 a day, let alone $1 a day, would starve and die. And many poor people in the world do. But those who do not die have more than their identifiable money income. They have the support which comes from networks of solidarity, enabling some people to survive severe and prolonged income poverty. The solidarity can take the form of the welfare state, the family or much looser but equally important forms of social cooperation.

■ World Bank 2000c.

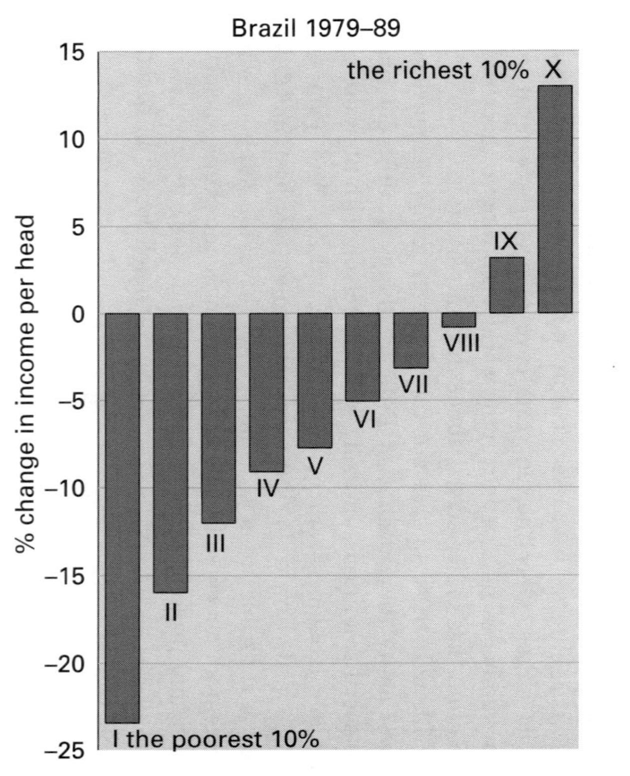
Brazil 1979–89
% change in income per head
15
10
5
0
–5
–10
–15
–20
–25
the richest 10% X
IX
VIII
VII
VI
V
IV
III
II
I the poorest 10%

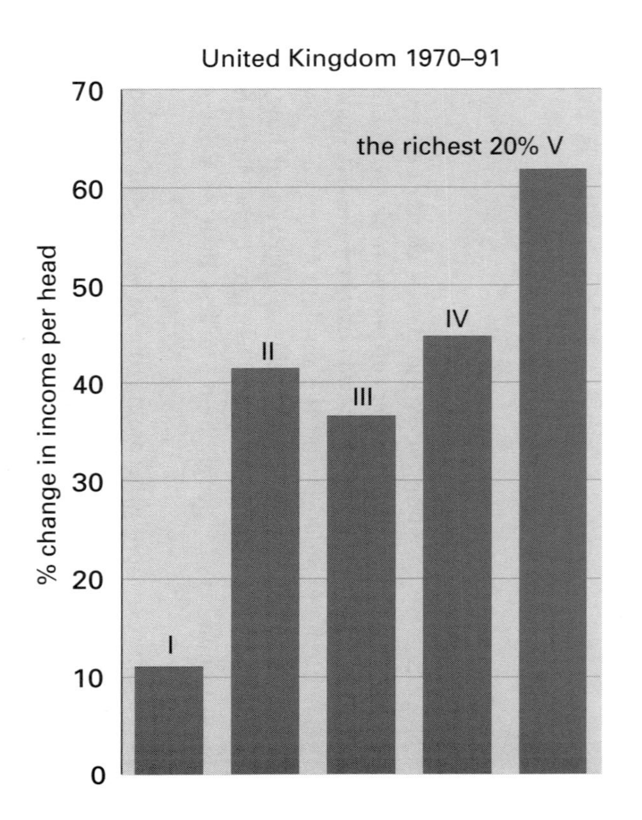
United Kingdom 1970–91
% change in income per head
70
60
50
40
30
20
10
0
the richest 20% V
IV
III
II
I

● The charts show the change in income for different income groups during the stated periods; the poorest group is on the left and the richest on the right. For Brazil the population is divided into ten groups (deciles) and for the UK into five (quintiles).

○ Both these charts show what happens during a period of growing inequality. In Brazil, as in much of Latin America, the decade of the 1980s is known as the 'lost decade' because it was the time of the debt crisis when countries repaid gigantic amounts of debt (› 86, 87), when economic growth slowed down or went into reverse and when programmes which benefited the poor were partly dismantled. But not everyone lost. The left-hand chart shows that there was a statistically perfect correlation between the order of economic groups and the price they paid. The two richest groups got richer and the rest got poorer. And the degree to which this happened depended exactly on the position of the group in the hierarchy.

In the UK the two decades after 1970 include both the effects of the first oil shock in 1973, after which growth slowed down, and of the liberalization and austerity policies known as supply-side economics, neo-liberalism or Thatcherism introduced after 1979. Although the income of all groups rose, the relative change was not unlike Brazil, an almost perfect relationship between level and change in economic situation. The richest got much richer and the poorest got only marginally less poor.

■ IABD 2000; World Bank 2000b; Deininger and Squire 1996.

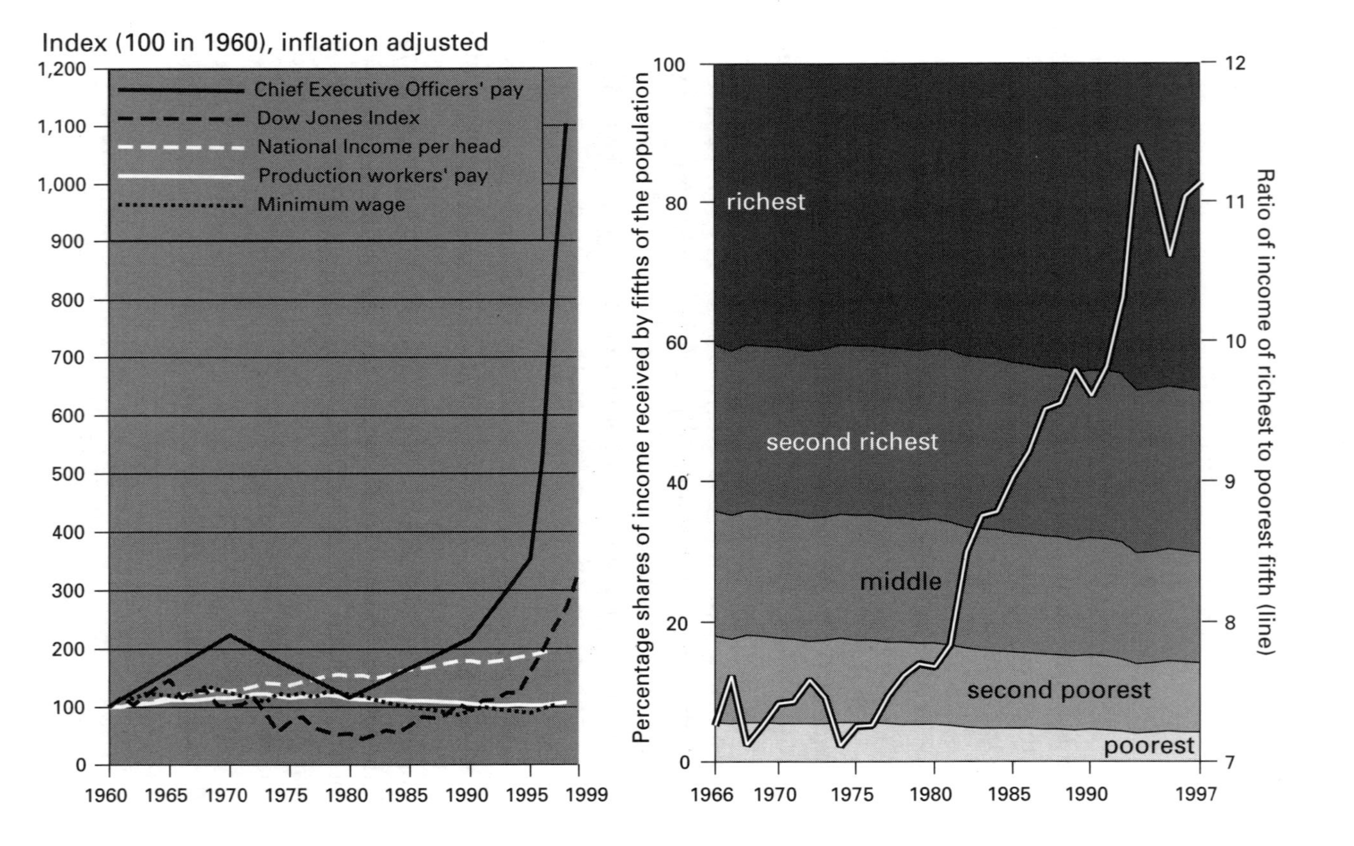

Index (100 in 1960), inflation adjusted
Chief Executive Officers' pay
Dow Jones Index
National Income per head
Production workers' pay
Minimum wage
1,200
1,100
1,000
900
800
700
600
500
400
300
200
100
0
1960
1965
1970
1975
1980
1985
1990
1995
1999
Percentage shares of income received by fifths of the population
Ratio of income of richest to poorest fifth (line)
richest
second richest
middle
second poorest
poorest
100
80
60
40
20
0
12
11
10
9
8
7
1966
1970
1975
1980
1985
1990
1997

The growth of inequality in the USA, 1960–1997

● The left-hand chart shows the movement between 1960 and 1999 of indices of five different sources of income in the USA. All start at 100 in 1960 so they compare changes in income, not levels of income. The right-hand chart shows changes in the distribution of US income from 1966 to 1997. The layers show how income is divided between fifths (quintiles) of the population; the line (measured against the right-hand axis) shows the ratio between the richest and the poorest fifth.

○ These charts tell a story of strongly growing inequality in the USA during recent decades. From the left-hand chart it can be seen that, during the last forty years, in which the average national income in real terms has doubled, the level of a production worker's pay, and of the minimum wage, have hardly changed at all. During the same period the real value of stock prices (as measured by the Dow Jones index), which partly determine the income of wealth owners, has increased by a multiple of more than 3. And, even more extraordinary, the salaries received by the chief executive officers of companies has expanded by a multiple of 11. This has been a polarization centred on differences of class. The owners and controllers of capital have reaped a bonanza while those who depend on wages or benefits have gained little or nothing.

The right-hand chart also shows the enormous acceleration of inequality after 1975. The line indicating the ratio of the income of the richest to the poorest fifth of the population rises from a value of about 7 in the mid-1970s to about 11 in the mid-1990s. And the five bands showing the division of income between the quintiles also clearly reveal how the rich have gained at the expense of the poor (‹10).

■ EPI 1999; EPI 2000; Global Financial Data 2000.

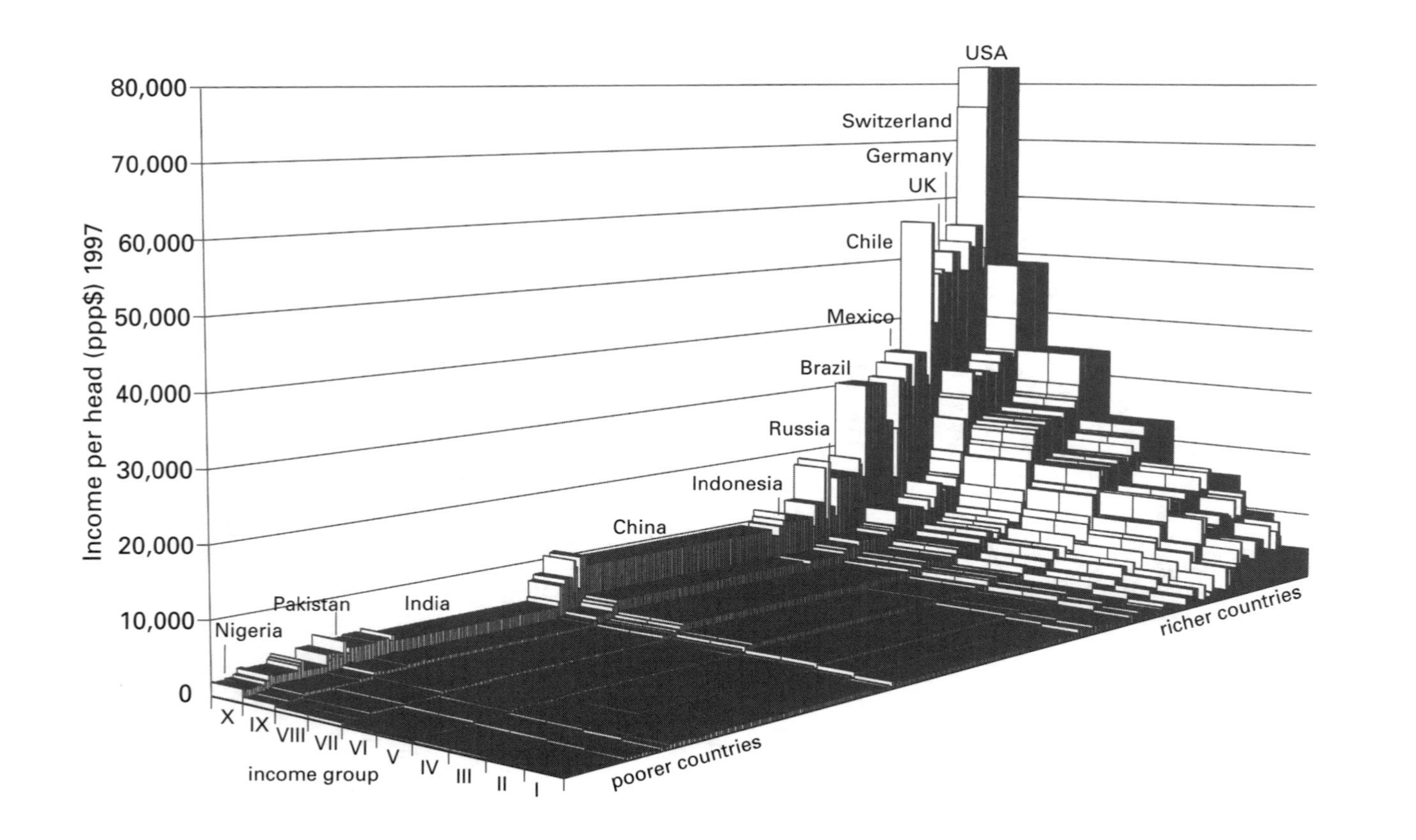
USA
Switzerland
Germany
UK
Chile
Mexico
Brazil
Russia
Indonesia
China
India
Pakistan
Nigeria
80,000
70,000
60,000
50,000
40,000
30,000
20,000
10,000
0
Income per head (ppp$) 1997
X
IX
VIII
VII
VI
V
IV
III
II
I
income group
poorer countries
richer countries

World distribution I: the unequal city

● Figures for the average income level of each country (in comparable **PPP** dollars) can be combined with information about the share received by each tenth (decile) of the population to calculate the implicit income of each group. This was already done in graphs 10, 11 and 13 for the richest and poorest groups. Here it is done for all groups and these are then viewed as a three-dimensional chart in which the height of the bars represents the income level of each group. From the left to the right we move from the poorer to the richer countries; from the back to the front we move from richer to poorer groups (tenths) within each national population. Each country has a number of these back-to-front rows of bars equivalent to its population (one row for every 10 million people). The result is an image of the distribution of income among the population of the whole world.

○ This striking image shows the world distribution of income as a kind of city in which each city block represents a tenth of the population of a country and the height of the building on the block is their average income. Most prominently in the far corner we see the skyscraper, or more appropriately Trump Tower, which represents the income of the richest tenth of the US population. Much less easy to decipher are the blocks of low-level buildings representing the poorest groups of the poorest countries in the nearest corner. The image conveys how that world inequality is a combination of differences within and between countries.

■ World Bank 1999a.

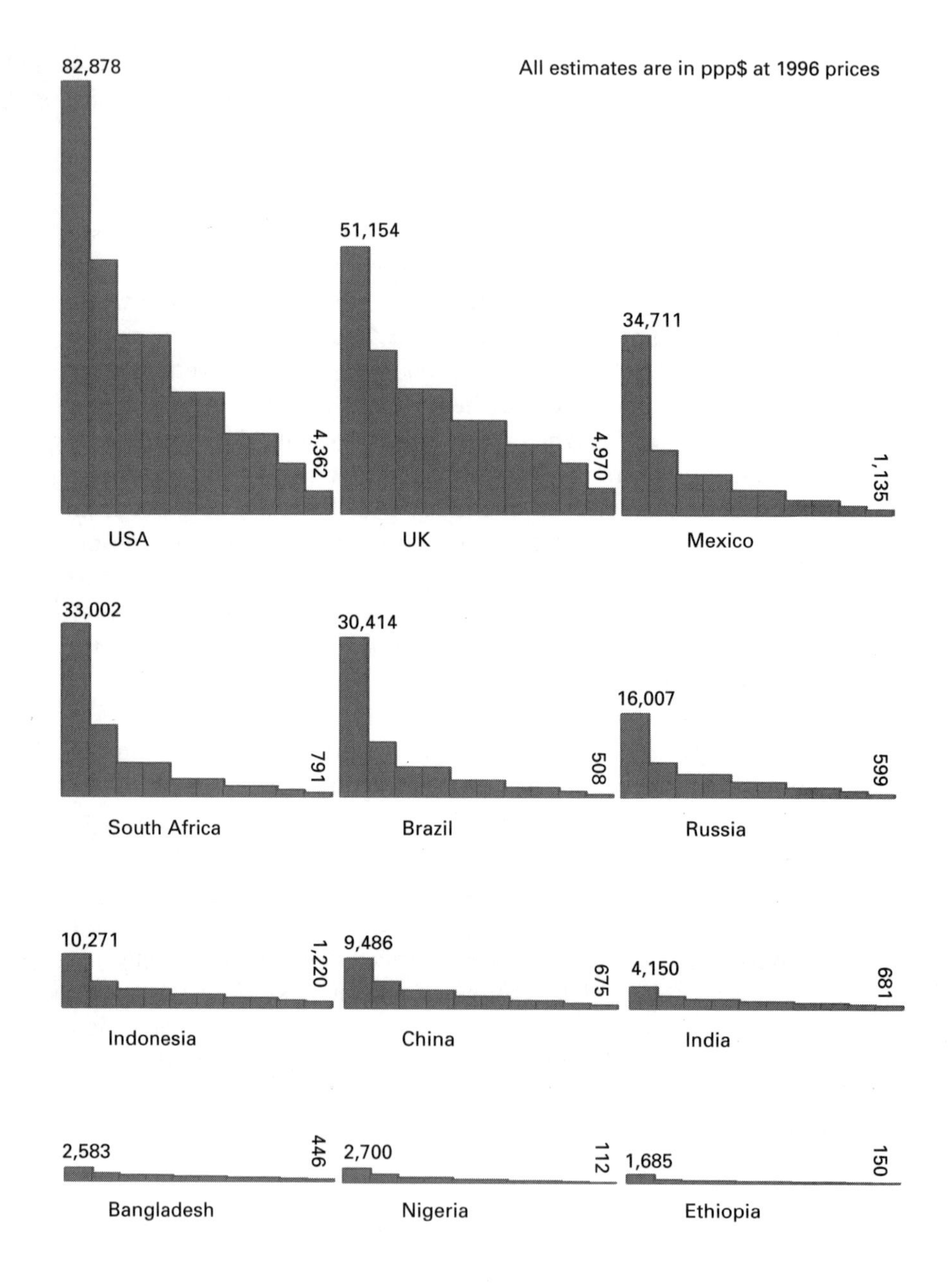
82,878
All estimates are in ppp$ at 1996 prices
51,154
34,711
4,362
4,970
1,135
USA
UK
Mexico
33,002
30,414
16,007
791
508
599
South Africa
Brazil
Russia
10,271
1,220
9,486
675
4,150
681
Indonesia
China
India
2,583
446
2,700
112
1,685
150
Bangladesh
Nigeria
Ethiopia

● If the previous chart was dismantled into its component parts they would look like this. These slices show the distribution of income for twelve sample countries with the level of income in dollars of the richest and the poorest tenths.

○ This gives another way of seeing a sample of the same information that is shown in the previous graph. Dismantling the graph draws attention to the differences in the degree of inequality between different countries. For this selection of countries the ratio of the income of the richest and the poorest tenths varies between three countries with ratios of over 30 to 1 (Mexico, South Africa and Brazil) and three with ratios of less than 10 (Indonesia, India and Bangladesh) (‹10).

The international difference between the rich is quantitatively greater than the international difference between the poor. The richest group in the USA has 49 times more than the richest group in Ethiopia, while the ratio between the poorest groups of the same two countries is 29 to 1.

These figures are again all in principle reduced to a currency unit which reflects parity of purchasing power. But they still have major limitations. It may be meaningful that the poorest tenth of the US population can buy a similar amount of commodities to the richest tenth of Indians. But their social position could hardly be further apart. There are many other dimensions to inequality than such material measures as this.

Even so it is still an important and chilling fact about the state of the world that one average member of the richest tenth of US citizens receives nearly as much as 750 of the poorest Nigerians.

■ World Bank 1999a.

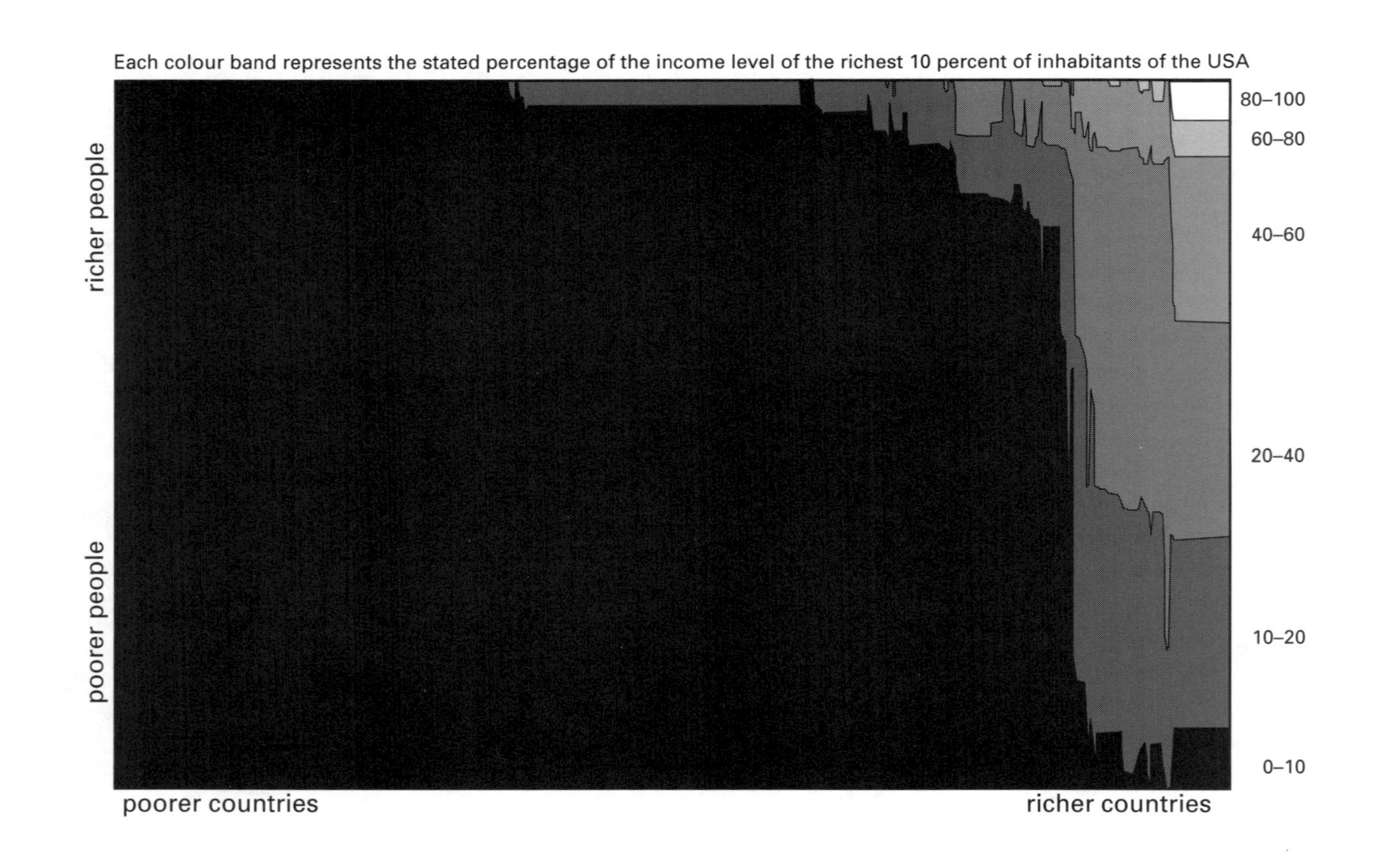
Each colour band represents the stated percentage of the income level of the richest 10 percent of inhabitants of the USA
80–100
60–80
40–60
20–40
10–20
0–10
richer people
poorer people
poorer countries
richer countries

World distribution II: unequal terrain

● This is exactly the same information as that shown in graph 13 but viewed in a different way. It is a kind of contour map of the distribution of world income. The whole space represents the population arrayed from left to right in order of the income of the country (poorest to richest) and from top to bottom in order of income within the country (richest to poorest). The bands display successive levels of income measured as a percentage of the income of the richest tenth in the USA. It is as if a huge cloth had been draped over the city in graph 13 and a contour map of it drawn.

○ This graph can be thought of as an area representing the population of the world: from the poorest people of the poorest countries in the bottom left-hand corner to the richest people in the richest countries in the top right-hand corner. A tiny proportion of the world population (coloured white in the top right-hand corner) receive the highest band of income (80–100 per cent of the income level of the richest US citizens). The enormous majority of the world's population (coloured black) receive less than 10 per cent of the reference level. As we move from left to right, from the poorer to the richer countries, the steepness of the contour is an indication of the very abrupt change between these two groups of countries. There are very few intermediate countries; there is a quite sharp gap in the distribution between North and South. This persistent fact provides considerable justification for looking at the world in terms of this binary division. Divided in this way, most of the intermediate population of the world (the various shades of grey) are the middle and poorer classes of the developed countries; a smaller number of them are the richer classes of under-developed countries.

■ World Bank 1999a.

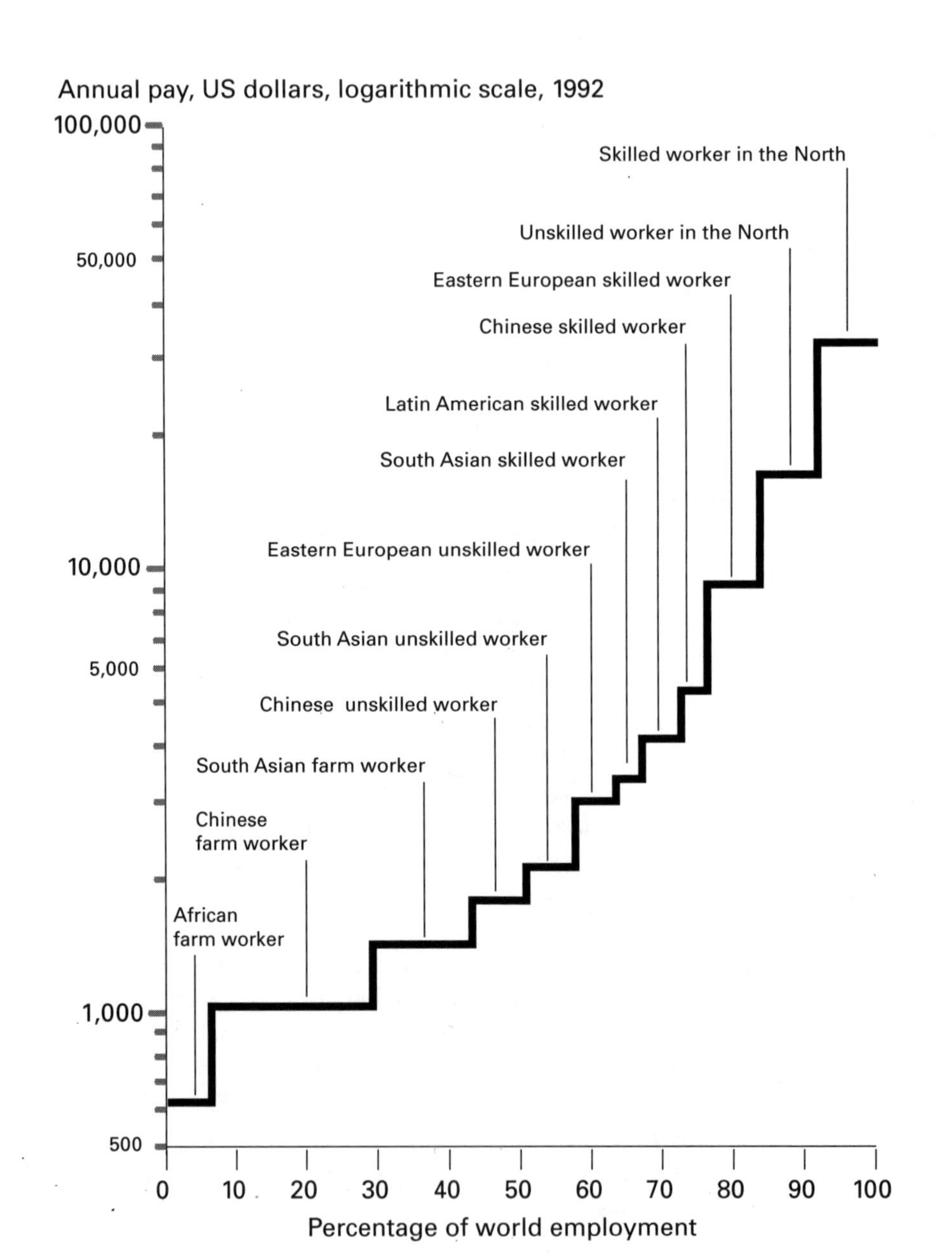

Annual pay, US dollars, logarithmic scale, 1992
100,000
50,000
10,000
5,000
1,000
500
Skilled worker in the North
Unskilled worker in the North
Eastern European skilled worker
Chinese skilled worker
Latin American skilled worker
South Asian skilled worker
Eastern European unskilled worker
South Asian unskilled worker
Chinese unskilled worker
South Asian farm worker
Chinese farm worker
African farm worker
0
10
20
30
40
50
60
70
80
90
100
Percentage of world employment

● The level of each step in this global wage staircase shows the average annual pay for the kind of worker described; the width of each step represents the percentage that the kind of worker described represents in the world paid labour force. In order to see such great differences in pay levels the scale is logarithmic: as the pay level increases the numbers are progressively more and more compressed. On a normal scale the staircase would appear much steeper.

○ This graph, compiled by the World Bank in 1995, shows the enormous variations in workers' pay according to three variables: geographical location, sector of work and level of skill. There are no surprises. Pay is usually higher for industrial and service workers than for agricultural workers, for workers in the North and for workers with skills. The actual hierarchy of workers' pay is in fact more complex than shown in this simple graph. The gap between workers has been growing in recent decades within countries of the North. This is partly associated with the kind of jobs that the new service economies create: a few very highly paid and many very low-paid personal service jobs. It is also associated with the growth of immigration and the frequent ghettoization of immigrant workers into the low-paying sectors. Some of the staircase of workers' inequality on a world scale, therefore, is increasingly appearing within national labour markets.

■ World Bank 1995a.

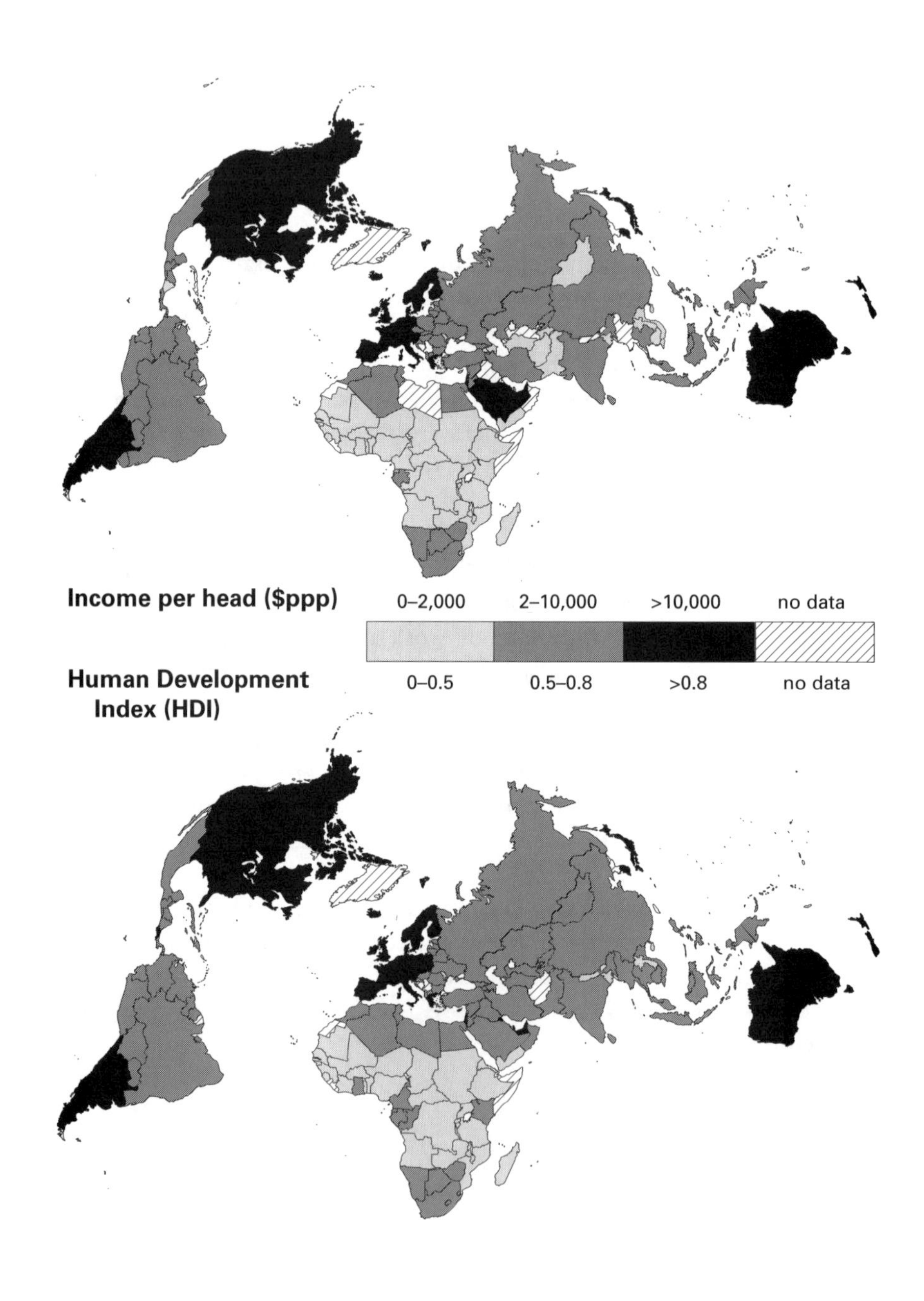

Income per head ($ppp)
0–2,000
2–10,000
>10,000
no data
Human Development Index (HDI)
0–0.5
0.5–0.8
>0.8
no data

Income and human development compared

● The lower map divides countries into three categories according to the value they are assigned in the UNDP's Human Development Index. The upper map (a simplified version of the one already seen in graph 8) divides countries by income per head converted to dollars by **purchasing power parity** estimates.

○ Up to here the graphs have mainly been about economic differences. This graph forms a bridge to the use of other less economic indicators of comparison. In 1990 the UNDP, with its Human Development Report, launched an attempt to dethrone income per head as the main measure of development. It recommended seeing development in a much broader way as an increase in human capacity and it proposed to measure it differently as well. Its Human Development Index (which can have a minimum value of 0 and a maximum of 1) is a complex average of three indicators or sets of indicators:

- national income per head attenuated at higher levels so that countries should not be rewarded simply for quantitative economic growth;
- a measure of health (life expectancy at birth);
- a measure of cultural development (an amalgamated index of adult literacy and the percentage of young people enrolled in primary, secondary and tertiary education).

These three components are given equal weight in calculating the overall index.

The UNDP has stressed that higher income is neither a necessary nor a sufficient condition for human development, the level of which tends to be determined more than anything else by enlightened government policies that give priority to improving health and education.

■ UNDP 2000.

Inequalities of births, lives, health and deaths

III

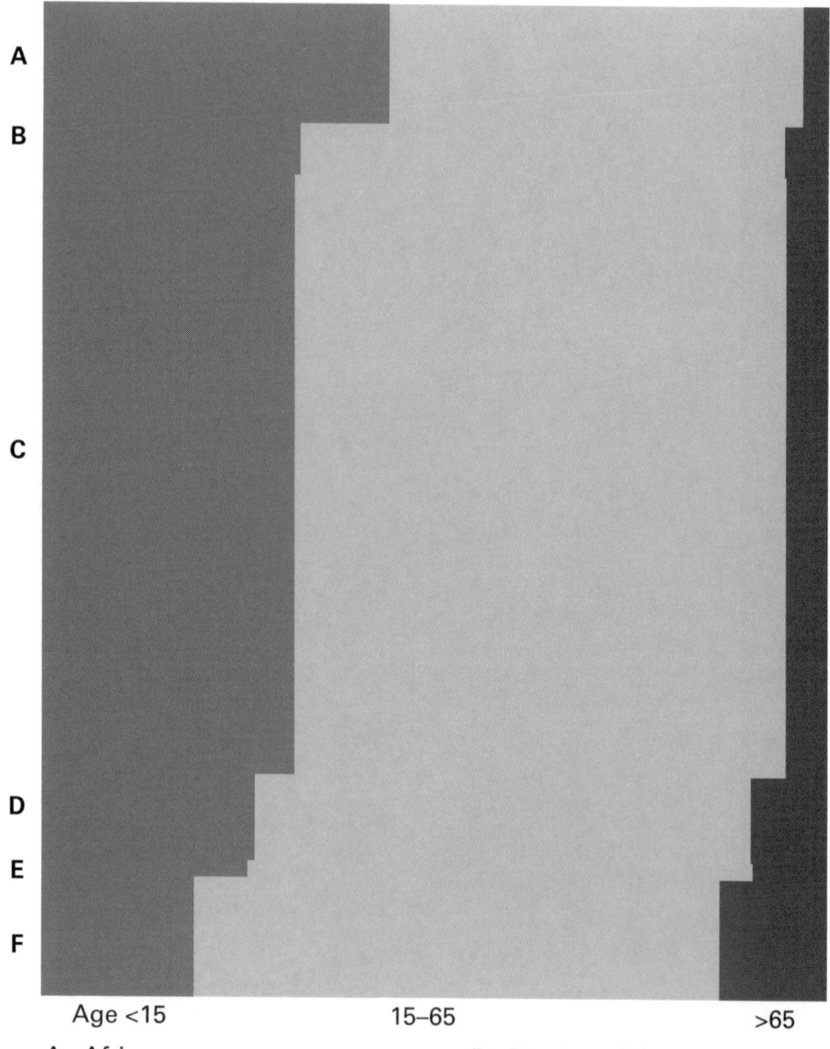
World total, 12 October1999: 6 billion
A
B
C
D
E
F
Age <15
15–65
>65
A Africa
B South America
C Asia
D North and Central America
E Oceania
F Europe

● The total area of the rectangle represents the world population. It is divided into horizontal bands of varying width according to the population of the region indicated and vertical bands (shaded differently) according to three age groups.

○ According to the demographers of the United Nations the world's population passed 6 billion in October 1999. This meant that it had grown by four times during the twentieth century, a rate of about 1.6 per cent a year. Since the beginning of human life the average growth rate has been in the order of 0.002 per cent.

The rapid growth of the population has been due not to any increase in the birth rate but to a virtually worldwide decrease in the death rate. Where this happened first, in the industrialized countries, birth rates have also fallen dramatically. The result of these changes is that there are proportionally more older people and fewer children in the population – in other words, population ageing. The graph shows this very clearly. The proportion of people between 15 and 64 years of age is very similar in all the regions but the proportions of old people and children vary inversely. As a result a very high proportion of the population of the developing countries is young. And an ever-growing proportion of the population of the developed countries is over 65.

The ratio between the population under 15 and over 65 to those aged 15–64 is often termed the dependency ratio. The economic and social appropriateness of this purely demographic concept is limited since some people between 16 and 64 are economic dependents while some below 15 and over 65 are not.

■ UN 1999.

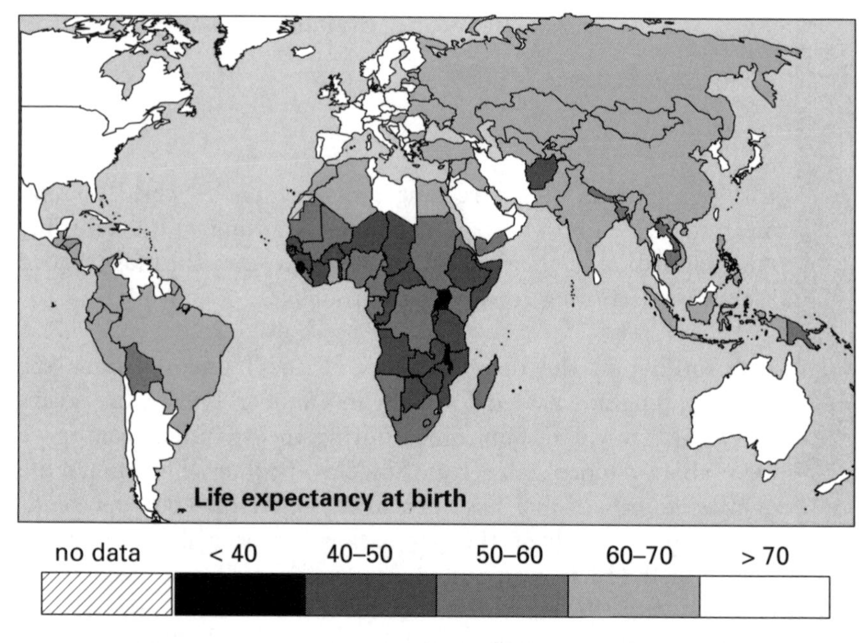
Life expectancy at birth
no data
< 40
40–50
50–60
60–70
> 70

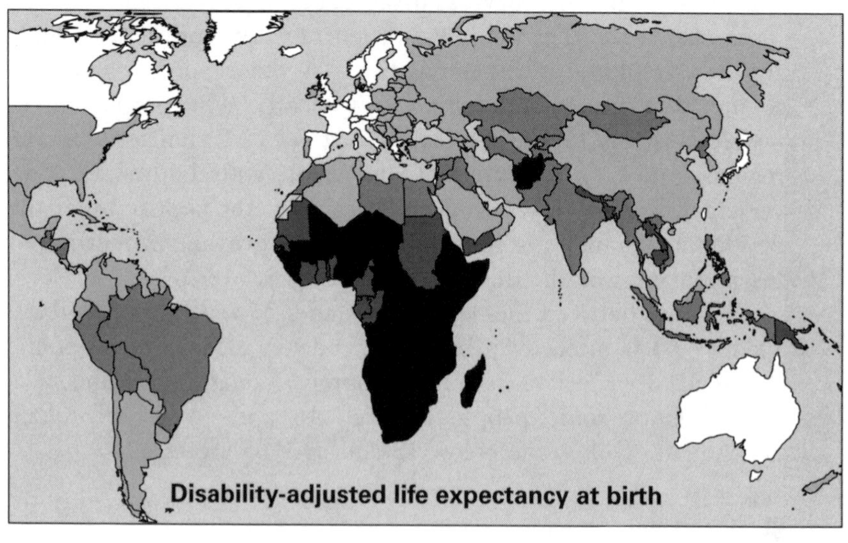
Disability-adjusted life expectancy at birth

● The upper map shows life expectancy at birth by country. The lower map shows disability-adjusted life expectancy.

○ Much attention is focused on the indicator life expectancy at birth. It is the number of years a person born today would most probably live if today's age-specific mortality continued into the future. If mortality rates continue to fall this indicator underestimates the years a person born today can expect to live. In the developed countries, and in a few countries of West Asia, North Africa and East Asia, as well as in a few Latin American countries, people on average can now, for the first time, expect to survive beyond the biblically allotted three score years and ten. At the other extreme in a small number of countries, mostly in Africa, people can only expect to survive about forty years, or even fewer.

This widely used indicator represents one-third of the Index of Human Development. It is routinely calculated for nearly all countries and is closely influenced by many other indicators which are more difficult to measure: for instance, the level of nutrition and access to and quality of medical services.

During the twentieth century life expectancy has risen in most countries with usually temporary reversals due to major wars and famines and epidemics such as the present AIDS epidemic (›120).

The World Health Organization now also calculates an alternative measure, disability-adjusted life expectancy (DALE), which reduces the value of the indicator in accordance with a calculation of expected illness and disability. This new index is more unequally distributed between countries than unadjusted life expectancy (›39).

■ UNDP 1999; WHO 2000.

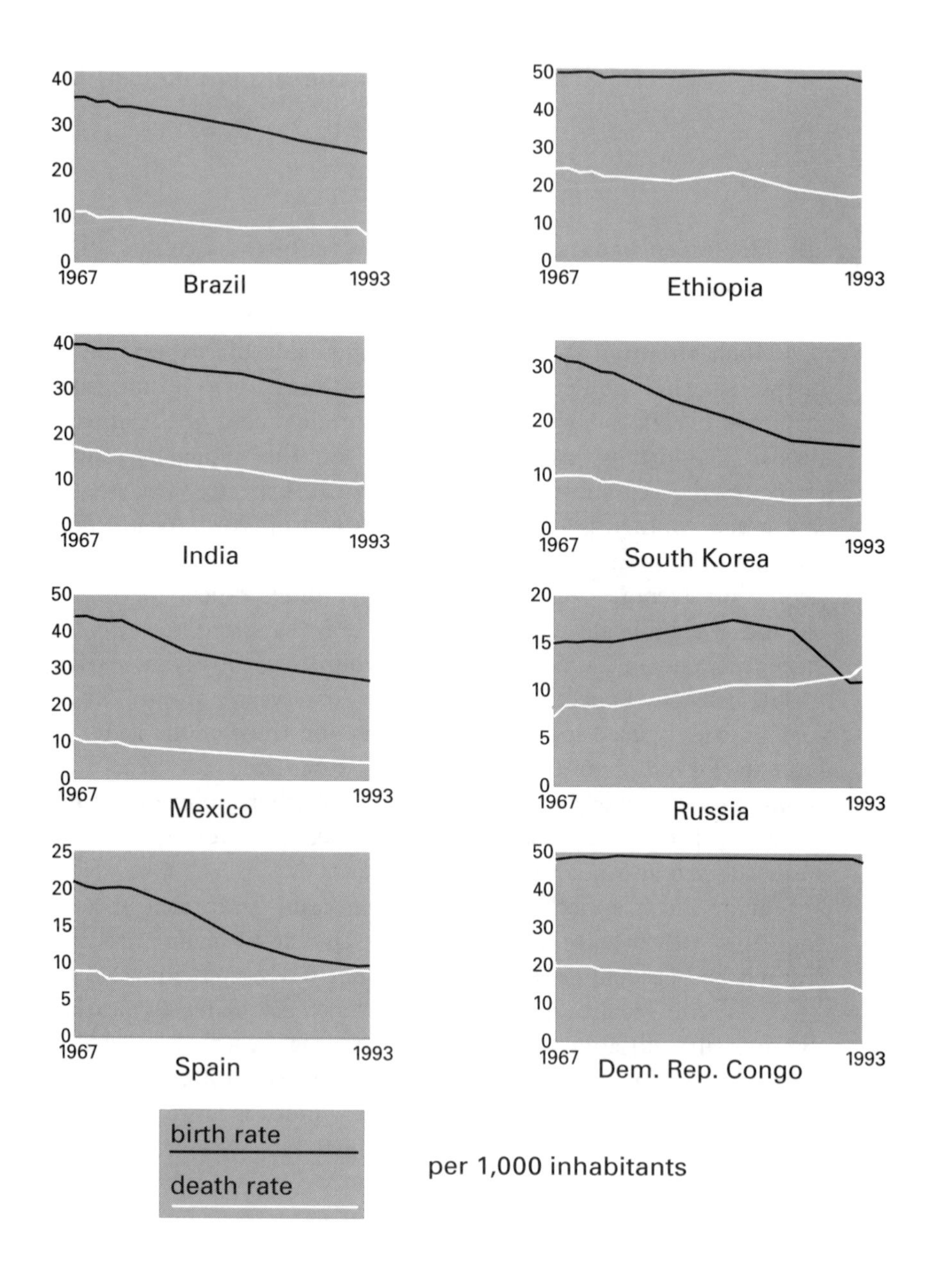

40
30
20
10
0
1967
Brazil
1993
50
40
30
20
10
0
1967
Ethiopia
1993
40
30
20
10
0
1967
India
1993
30
20
10
0
1967
South Korea
1993
50
40
30
20
10
0
1967
Mexico
1993
20
15
10
5
0
1967
Russia
1993
25
20
15
10
5
0
1967
Spain
1993
50
40
30
20
10
0
1967
Dem. Rep. Congo
1993
birth rate
death rate
per 1,000 inhabitants

Demographic transitions, 1967–1993

● The graphs show the evolution of the overall birth and death rates in eight sample countries between 1967 and 1993.

○ The recent experience of these eight countries shows how general is the decline in death rates mentioned earlier; but it also shows that the pattern of demographic change is far from uniform. Brazil, India, Mexico and South Korea conform to the expected model: falling death rates followed by falling birth rates, leading albeit slowly to a decline in population growth. This process is known as the 'demographic transition'. But in the two African countries, Ethiopia and the Democratic Republic of the Congo, the birth rate has hardly fallen despite a decline in death rates. Russia is in all respects a special case. A serious mortality crisis, the causes of which are still being discussed, has existed in the country from before the collapse of communist government; and a sharp decline in the birth rate has thrown population growth into reverse. And Spain is another special case, an extreme version of what is happening in most European countries. The death rate has reached a minimum and is slowly climbing again due to the ageing of the population. And that is produced not only by greater longevity but also by a sharp decline in the birth rate. In the near future, on present trends, birth rates throughout Europe will be less than death rates and the population will fall substantially.

It is now widely debated whether depopulation and ageing are inevitable and whether they are undesirable side effects of the benefits of better nutrition, medical improvements and greater choice about childbirth.

■ UN 1999.

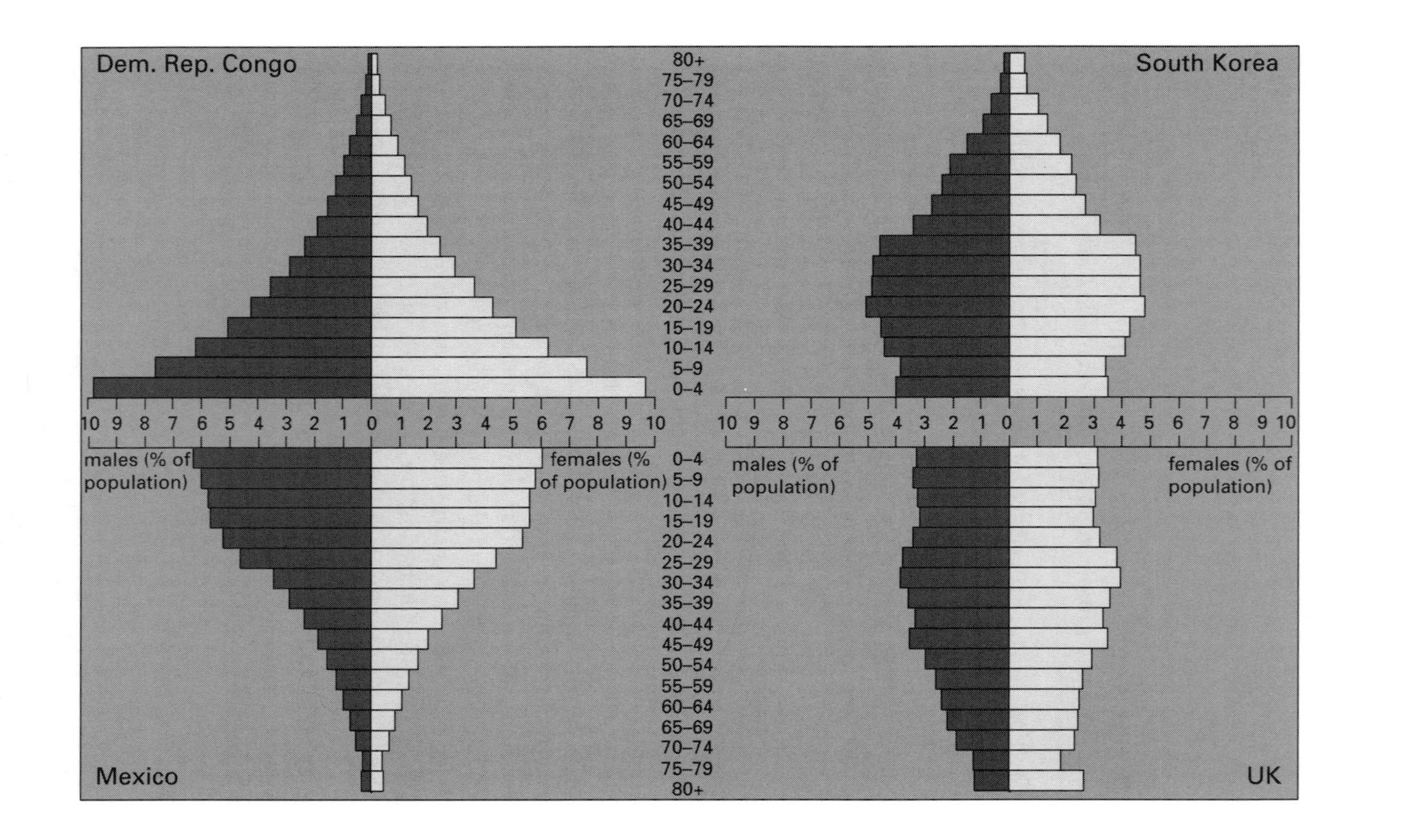
Dem. Rep. Congo
South Korea
Mexico
UK
80+
75–79
70–74
65–69
60–64
55–59
50–54
45–49
40–44
35–39
30–34
25–29
20–24
15–19
10–14
5–9
0–4
10 9 8 7 6 5 4 3 2 1 0 1 2 3 4 5 6 7 8 9 10
males (% of population)
females (% of population)

● The population pyramids show the population by age group and sex, measured as a percentage of the total population of the country in 1995.

○ These pyramids give a very vivid picture of the enormous change in human society produced by demographic change. Before any demographic transition has taken place the population of each age group is greater than the one above it. High birth and death rates maintain the broad-based pyramid shape and stability of total numbers. Initially the fall in the death rate, especially child mortality, broadens the base of the pyramid and accelerates population growth. Finally as birth rates fall the pyramid narrows at the base and becomes a tower (as in the case of the UK) or even in more extreme cases an inverted pyramid and population once again becomes stable or even falls.

The age structure of populations is one of the main causes of difference between national human societies. It changes the type of needs and priorities that each country has. For countries with high proportions of young people the provision of education and of children's health, and then of employment for a rapidly growing labour force, are crucial. In a more aged society with a low birth rate much more importance is attached to old people's health, to pensions and how to pay them. Most changing societies have found that they have entered into the second set of concerns before they have fully solved the first. And the problems associated with ageing, even though it is completely predictable, catch most societies unprepared.

■ UN 1999.

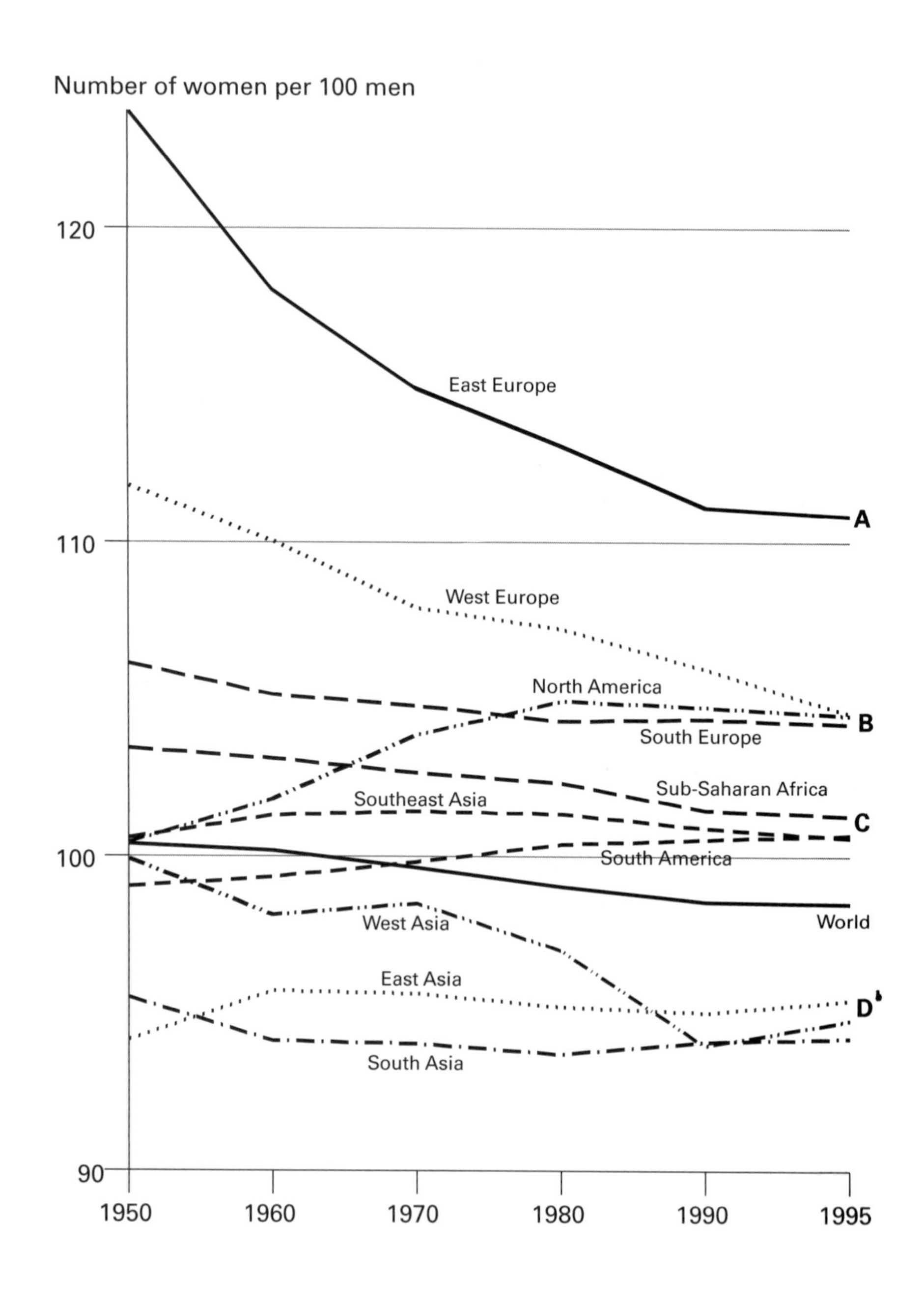

Number of women per 100 men
120
110
100
90
1950
1960
1970
1980
1990
1995
East Europe
A
West Europe
North America
B
South Europe
Sub-Saharan Africa
Southeast Asia
C
South America
World
West Asia
East Asia
D
South Asia

● The lines show the female population as a percentage of the male population, for the world and by major region, during the last half-century. A figure above 100 means that women are the majority of the population; under 100 and they are in the minority.

○ It is widely (and rightly) believed that, for naturally or socially determined causes, women tend to live longer than men. So, it seems that ageing must be associated with an increasing proportion of women in the population. Yet probably the least known major demographic fact about the world is that since about 1965 men have formed the majority of the population, and their majority is slowly but steadily increasing. This is what the line for the world shows in this chart.

But the average of the world is not typical of any of its major regions. This chart suggests that there are four quite different levels for the ratio of women to men:

A In Eastern Europe there is a large majority of women, partly still due to the effects of excess male mortality in World War II, partly to a continuation of exceptional male mortality.

B In the developed countries there is a convergence at about 105 women to every 100 men, caused by population ageing, immigration patterns and the decline in the effects of World War II.

C In South America, Sub-Saharan Africa and Southeast Asia there is another convergence around a smaller predominance of women (about 102 to 100).

D In West, East and South Asia there is a clear majority of men. These are therefore the areas responsible for the world's male majority.

■ UN 1994; UN 1999.

Countries with lowest % of women (< 49%)

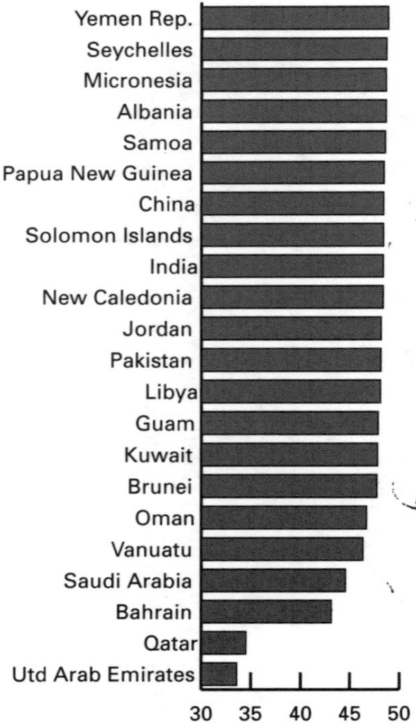

Yemen Rep.
Seychelles
Micronesia
Albania
Samoa
Papua New Guinea
China
Solomon Islands
India
New Caledonia
Jordan
Pakistan
Libya
Guam
Kuwait
Brunei
Oman
Vanuatu
Saudi Arabia
Bahrain
Qatar
Utd Arab Emirates

30 35 40 45 50

Shaded countries have male majorities

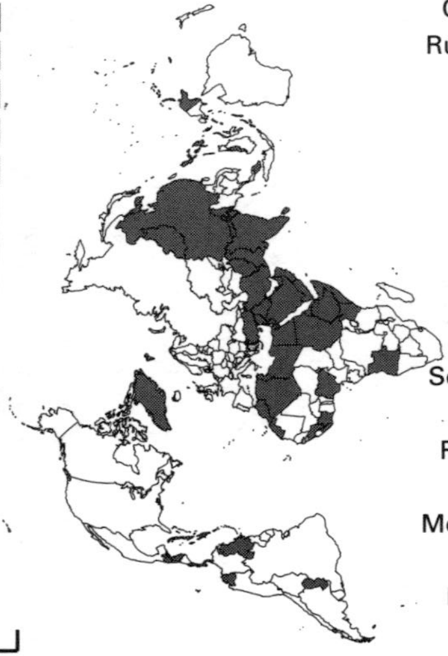

Countries with highest % of women (> 51.5%)

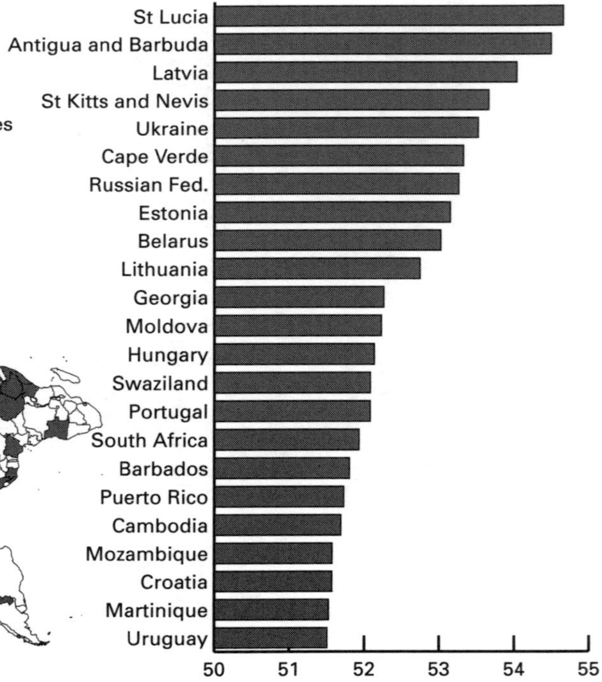

St Lucia
Antigua and Barbuda
Latvia
St Kitts and Nevis
Ukraine
Cape Verde
Russian Fed.
Estonia
Belarus
Lithuania
Georgia
Moldova
Hungary
Swaziland
Portugal
South Africa
Barbados
Puerto Rico
Cambodia
Mozambique
Croatia
Martinique
Uruguay

50 51 52 53 54 55

Differences in the proportion of men and women

● The chart on the left shows all the countries with the largest male majority in 1995, and the one on the right all the countries with the largest female majority. In the map the countries with a male majority are shaded.

○ The figures for the sexual division of national populations tell several stories. One is a story of immigrant labour. The oil-producing countries of the Gulf region have high proportions of men because their great recent wealth and low population have led them to rely to a great extent on immigrant labour from the rest of Asia; and most of these immigrants are men. At the other extreme some of the countries with the largest female majority demonstrate the counterpart: they are countries of emigration, like some Caribbean islands, where most emigrants are men. The second category of countries with a large female majority consists of virtually all the countries of the former USSR and other Eastern European countries. The huge deficit of men as a result of deaths, either at the hands of Hitler's invasion or of Stalin's repression, still persists among the older age groups. But the majority of women, though it has been declining, is not approaching the levels of the rest of Europe. A sudden increase in adult mortality during the 1990s affected men disproportionately and was closely associated with increased consumption of alcohol.

The countries that determine the world sexual balance of population are those large countries in Asia, especially the giants China and India, which both have male majorities.

■ UN 1994; UN 1995; UN 1999; World Bank 2000b.

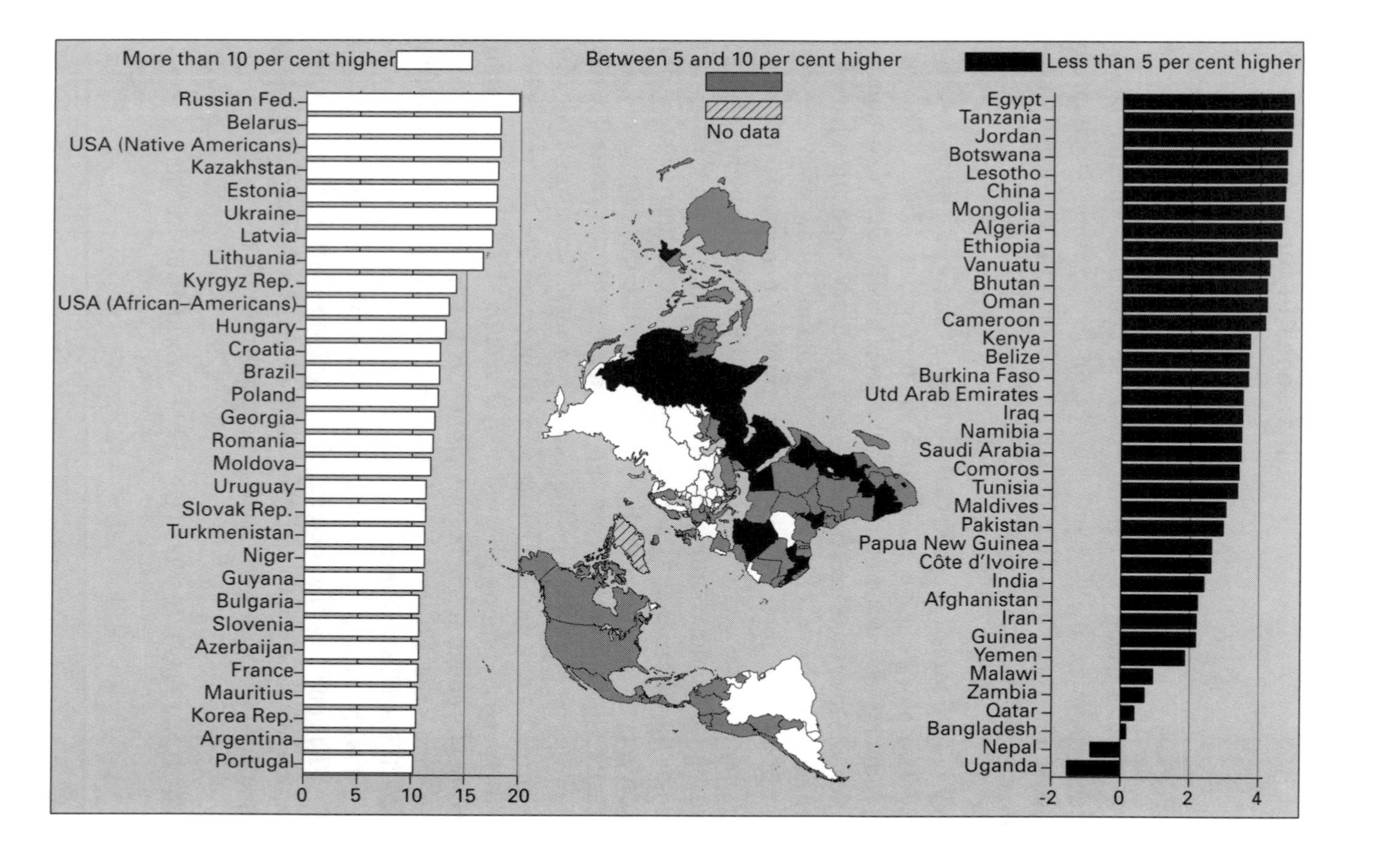
More than 10 per cent higher
Between 5 and 10 per cent higher
No data
Less than 5 per cent higher
Russian Fed.
Belarus
USA (Native Americans)
Kazakhstan
Estonia
Ukraine
Latvia
Lithuania
Kyrgyz Rep.
USA (African–Americans)
Hungary
Croatia
Brazil
Poland
Georgia
Romania
Moldova
Uruguay
Slovak Rep.
Turkmenistan
Niger
Guyana
Bulgaria
Slovenia
Azerbaijan
France
Mauritius
Korea Rep.
Argentina
Portugal
0
5
10
15
20
Egypt
Tanzania
Jordan
Botswana
Lesotho
China
Mongolia
Algeria
Ethiopia
Vanuatu
Bhutan
Oman
Cameroon
Kenya
Belize
Burkina Faso
Utd Arab Emirates
Iraq
Namibia
Saudi Arabia
Comoros
Tunisia
Maldives
Pakistan
Papua New Guinea
Côte d'Ivoire
India
Afghanistan
Iran
Guinea
Yemen
Malawi
Zambia
Qatar
Bangladesh
Nepal
Uganda
-2
0
2
4

Female compared with male life expectancy

● The left-hand chart shows all the countries where women's life expectancy in 1997 was 10 per cent or more higher than men's (coloured white on the map); and that on the left shows all the countries where it is less than 5 per cent higher than men's (coloured black on the map). In the majority of countries of the world the difference is between 5 and 10 per cent (coloured grey on the map).

○ In all countries of the world except two (Nepal and Uganda) women live longer than men. But in most of those countries that have a male majority women's superiority in life expectancy is less than 5 per cent. Also in most of the countries identified with especially large female majorities women's life expectancy is more than 10 per cent greater than men's.

As with all national statistics, countries are not necessarily homogeneous. But usually data by different groups or regions is more difficult to acquire. In this case, however, there is information about different ethnic communities in the USA. While in that country the overall national difference between male and female life expectancy falls in the range between 5 and 10 per cent, both Native American and African-American women live more than 10 per cent longer than the men of those communities. Demographically they are more like Russians or Ukrainians than US whites.

If almost everywhere women live longer than men, why are there in total fewer women than men? There is only one possible answer to that question and it is shown in the following chart.

■ UNDP 1999; Hacker 1995.

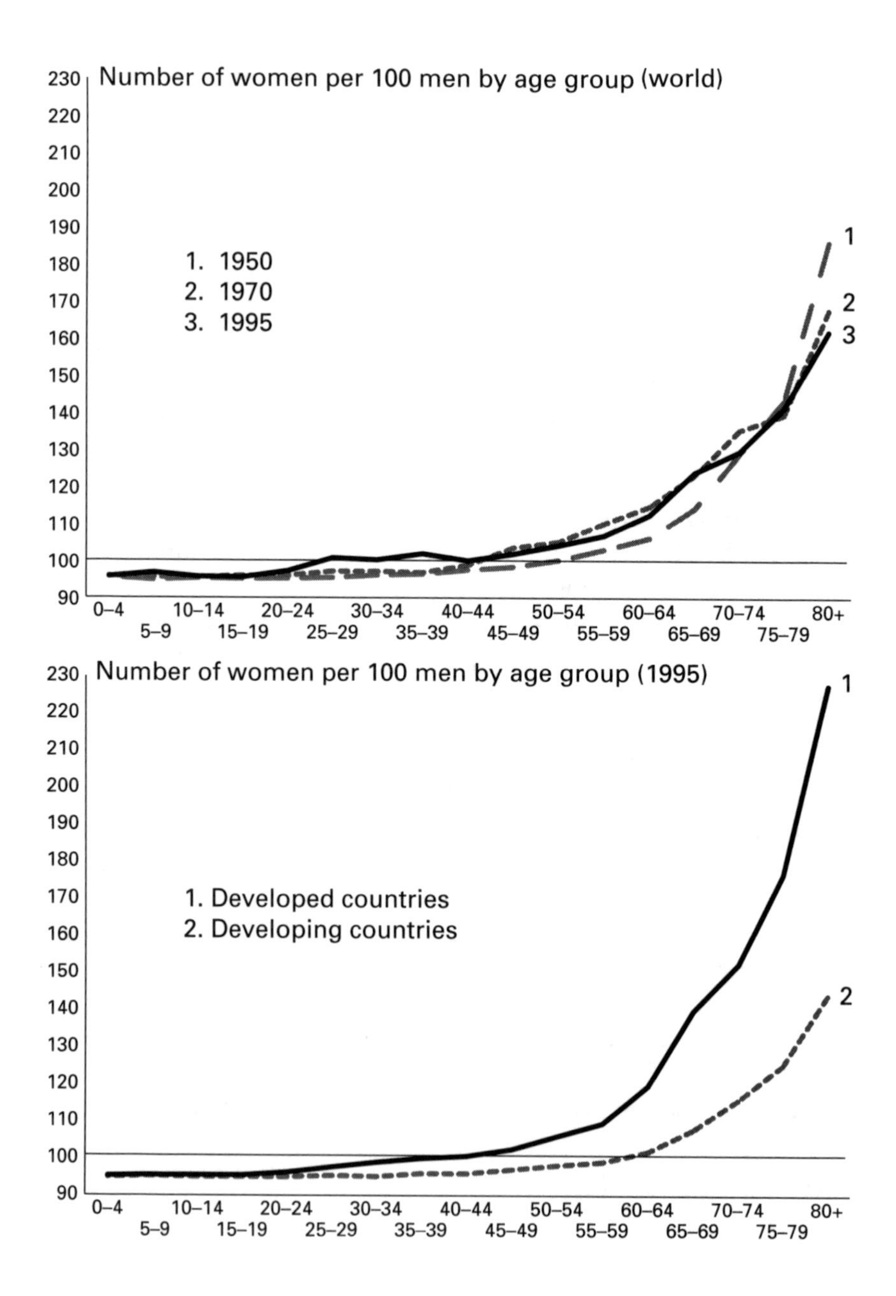

Number of women per 100 men by age group (world)
1. 1950
2. 1970
3. 1995
230
220
210
200
190
180
170
160
150
140
130
120
110
100
90
0–4
5–9
10–14
15–19
20–24
25–29
30–34
35–39
40–44
45–49
50–54
55–59
60–64
65–69
70–74
75–79
80+
1
2
3
Number of women per 100 men by age group (1995)
1. Developed countries
2. Developing countries
230
220
210
200
190
180
170
160
150
140
130
120
110
100
90
0–4
5–9
10–14
15–19
20–24
25–29
30–34
35–39
40–44
45–49
50–54
55–59
60–64
65–69
70–74
75–79
80+
1
2

Men and women: the effects of age, time and development

● The lines show the ratio of the numbers of women to men in the population, by age group. The upper chart is the world population in three different years. The lower chart shows the ratios for the developed and the developing countries separately in 1995.

○ There is a clear pattern to the sex ratio of the human population. More males than females are born and so in the early years there is a majority of men (fewer than 100 women for every 100 men). But male mortality is usually higher than female mortality and so the male majority is gradually eliminated in higher age groups. And in the oldest groups the superior longevity of women means that there is a growing and eventually very large majority of women. The lower chart shows that this effect is just as clear but considerably less strong in developing countries than in developed countries. This difference, as the upper chart shows, is the reason that the effect has become less marked during the last fifty years: the developing countries, with a male majority overall, have formed a progressively increasing part of the world population.

The lower chart shows that the change between a male majority at younger ages and a female majority at older ages takes place around 40 years old in the developed countries and around 60 years old in the developing ones. That difference, of course, does much to explain the world's overall male majority. Developing countries, as we have seen, have relatively fewer people in the older age group in which women tend to predominate and their share of the world's population has been growing.

■ UN 1994.

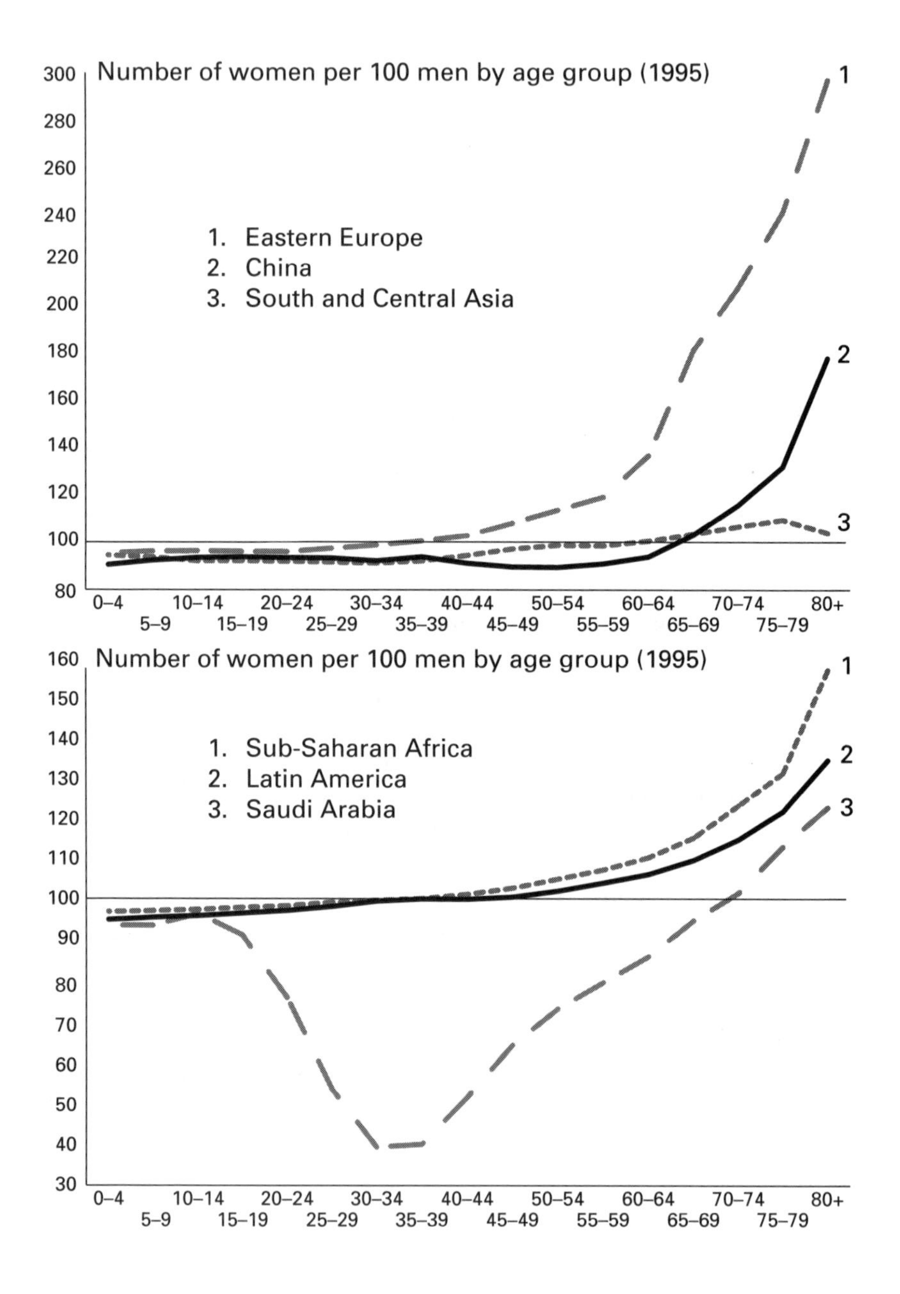
Number of women per 100 men by age group (1995)
1. Eastern Europe
2. China
3. South and Central Asia
300
280
260
240
220
200
180
160
140
120
100
80
0–4
5–9
10–14
15–19
20–24
25–29
30–34
35–39
40–44
45–49
50–54
55–59
60–64
65–69
70–74
75–79
80+
1
2
3
Number of women per 100 men by age group (1995)
1. Sub-Saharan Africa
2. Latin America
3. Saudi Arabia
160
150
140
130
120
110
100
90
80
70
60
50
40
30
0–4
5–9
10–14
15–19
20–24
25–29
30–34
35–39
40–44
45–49
50–54
55–59
60–64
65–69
70–74
75–79
80+
1
2
3

Men and women: the effects of geography and migration

● These lines show the same information as those on the previous page. In this case they are for six regions or countries in 1995.

○ The previous charts were for the world or large groups of countries. These show more of the specific characteristics of regions and countries and they are, therefore, more different from each other. In Eastern Europe (1 in the upper chart) excess male mortality means that women become the majority at about 35 years of age. By contrast, in China (2 in the upper chart) this does not take place until after 60 years of age, which explains the large male majority in that country. India (3 in the upper chart) shows a similar pattern to China up to 65 years of age. But then the two giant countries diverge completely. China displays the almost universal phenomenon of much greater longevity of women and so a large majority of women in older age groups; India does not. In other words in India older women's mortality is not less than men's.

In the lower chart Sub-Saharan Africa and Latin America (1 and 2) display a similar pattern: women become the majority beyond the age of 35, and for higher age groups the female majority grows considerably, though the effect is more marked in Latin America. In both these continents women are in the majority in the whole population. The case of Saudi Arabia (3) is very peculiar: the enormous increase in the male majority after about 15 years is the result of large-scale male immigrant labour. After 39 years the curve turns upwards and the male majority decreases due to a combination of the return home of the migrant workers and the greater mortality of the native male population. Among older people here, too, there is a majority of women.

■ UN 1994.

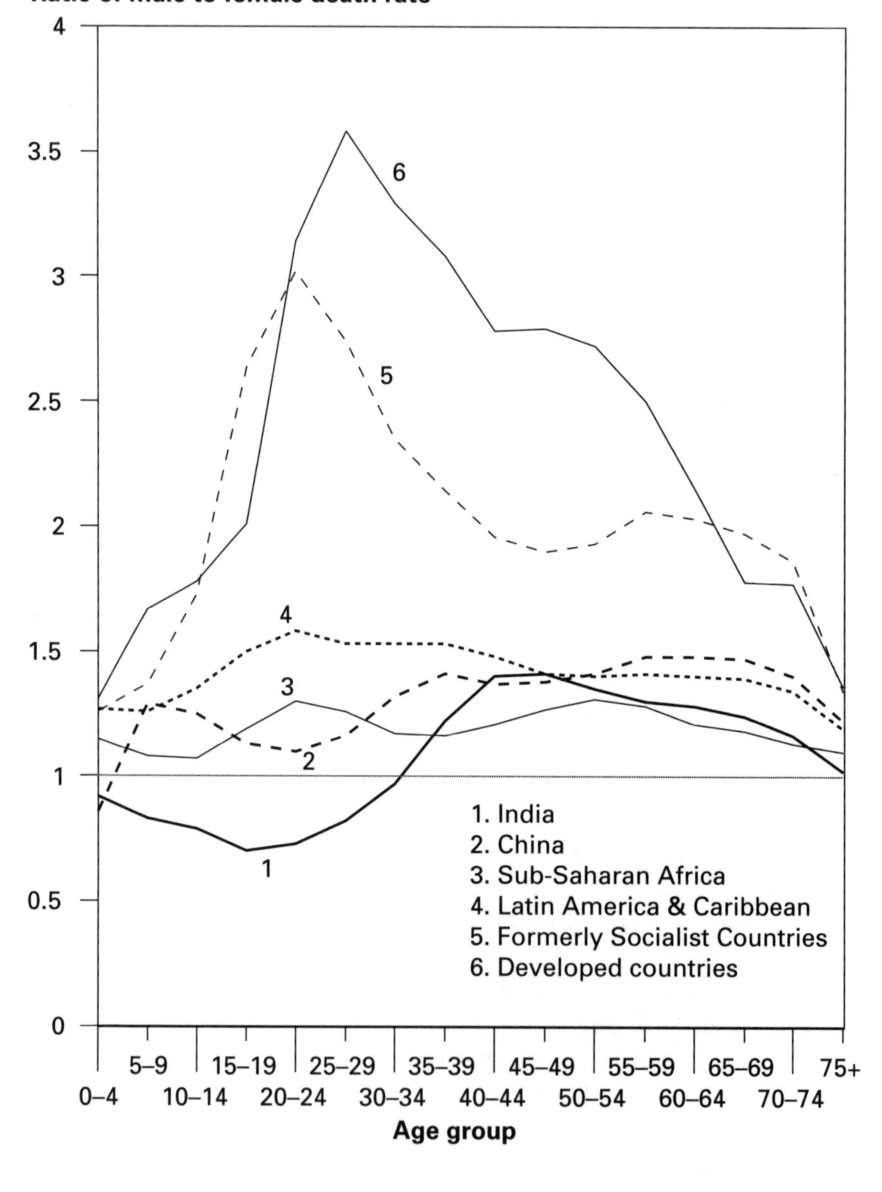

Ratio of male to female death rate
4
3.5
3
2.5
2
1.5
1
0.5
0
6
5
4
3
2
1
1. India
2. China
3. Sub-Saharan Africa
4. Latin America & Caribbean
5. Formerly Socialist Countries
6. Developed countries
0–4
5–9
10–14
15–19
20–24
25–29
30–34
35–39
40–44
45–49
50–54
55–59
60–64
65–69
70–74
75+
Age group

● The lines show, for the indicated regions, the ratio between the male and female mortality rate by age group.

○ Male mortality rates are slightly higher than female rates in all age groups in nearly all countries, leaving almost everywhere a majority of women in older age groups. It is important, therefore, to observe the exceptions. Of the six country categories compared here the one exception to the rule that male mortality is higher is in young age groups in India, another way of observing the phenomenon of missing women, already well known from the work of Amartya Sen and others. In India, unlike the other five regions, girls and young women die more than boys and young men, a fact which may be caused by discrimination of many kinds: in the diet, in access to medical treatment in the case of illness (›34) and in physical cruelty, including murder.

The other exceptional thing shown here is instances of exaggeratedly high male mortality. Young men in developed countries and former socialist countries have much higher mortality rates than young women. Between the ages of 20 and 30 male mortality is three to four times as great as female. The phenomenon is not seen in other parts of the world; nor in the historical statistics of England and Wales up to the Second World War. The 1938 UK ratios show a similar pattern to that of India today. By 1965, however, two peaks in relative male mortality, in youth and later middle age, had appeared. A feature of actually existing development and formerly existing socialism, it appears, is an extraordinary, and very recent, difference in the degree of risk for young men and women, influenced, one can surmise, by the fatal encounter between modern machismo and alcohol, drugs, firearms and motor vehicles.

■ Murray and Lopez 1996.

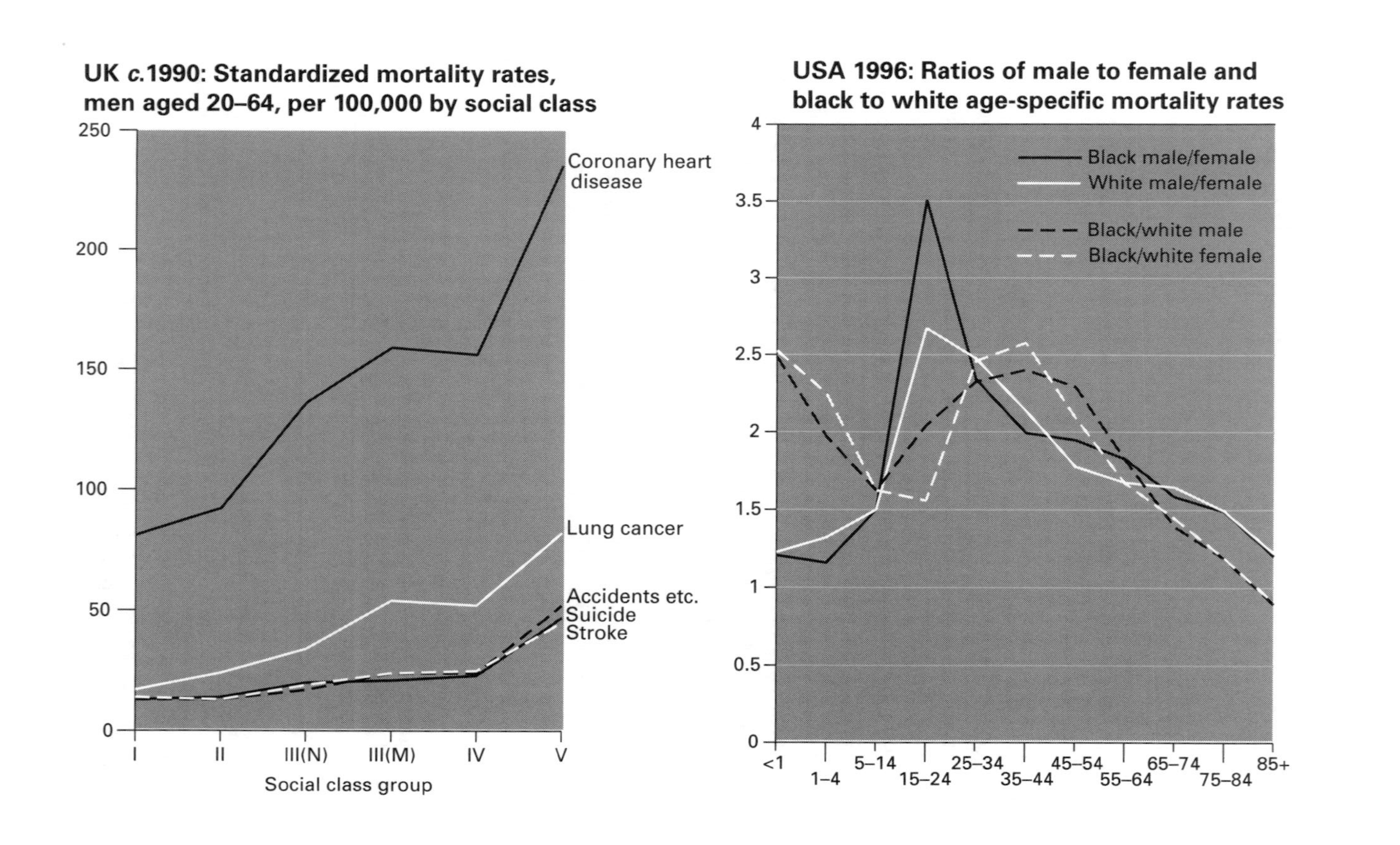

UK c.1990: Standardized mortality rates, men aged 20–64, per 100,000 by social class
250
200
150
100
50
0
Coronary heart disease
Lung cancer
Accidents etc.
Suicide
Stroke
I
II
III(N)
III(M)
IV
V
Social class group
USA 1996: Ratios of male to female and black to white age-specific mortality rates
4
3.5
3
2.5
2
1.5
1
0.5
0
Black male/female
White male/female
Black/white male
Black/white female
<1
1–4
5–14
15–24
25–34
35–44
45–54
55–64
65–74
75–84
85+

● The left-hand chart shows the mortality rate by social class (I, upper professional, to V, low-paid manual worker) for five different fatal conditions in the UK around 1990. The right-hand graph refers to the USA in 1996 and shows four ratios of mortality rates by age: men/women for whites and blacks, and white/black for men and women.

○ Ethnic difference and inequality are other fundamental determinants of mortality differences. In South Africa, the USA and Brazil whites live between seven and ten years longer than blacks or non-whites. The right-hand chart shows, for the USA, the ratio of age-specific mortality rates for males (black/white) and for females (black/white) as well as the male/female ratios for blacks and whites. These data again show an extraordinary peak in male mortality in the 15–24 age group common to both groups, though considerably more pronounced for blacks. For all age groups (except, curiously, the over 85s) the black mortality rate is higher than the white. The greatest differences are to be seen in child mortality and in the age groups from 25 to 55, where the difference is especially marked for women.

Mortality varies also with economic level or social class. Numerous studies, especially in the developed countries, have confirmed that the materially and socially privileged live longer and healthier lives than the poor. The left-hand chart shows how mortality rates from five causes among men in the UK are all closely connected with social class. It remains to be shown exactly what combination of detailed factors accounts for these facts – they could include worse diet, less access to medical care, and the health effects of stress factors associated with type of work and with relative poverty and economic uncertainty.

■ United States Census Bureau 1999; Acheson et al. 1998.

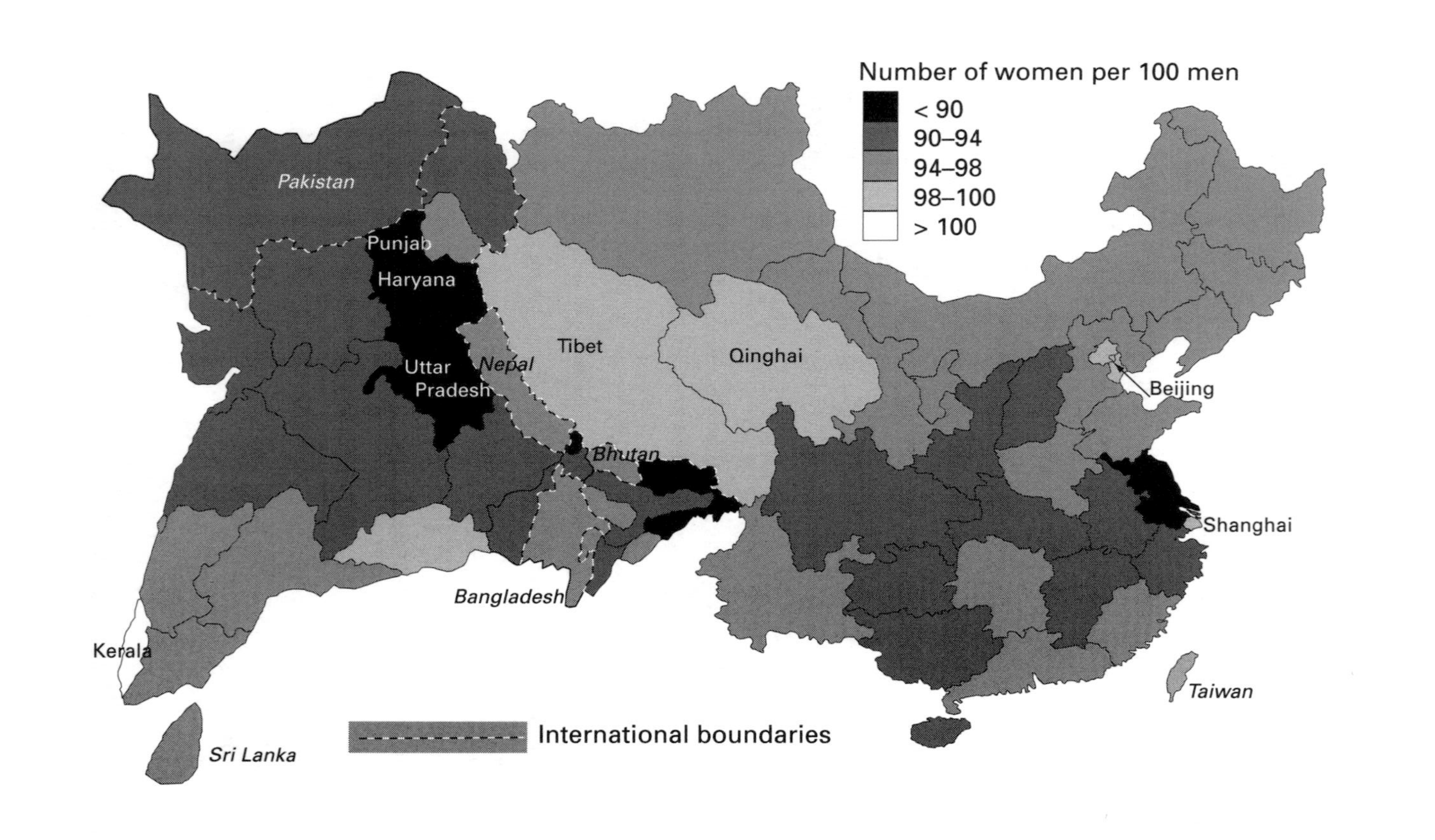
Number of women per 100 men
< 90
90–94
94–98
98–100
> 100
Pakistan
Punjab
Haryana
Uttar Pradesh
Nepal
Tibet
Qinghai
Bhutan
Bangladesh
Kerala
Sri Lanka
Beijing
Shanghai
Taiwan
International boundaries

The sex ratio in China and South Asia

● This map shows differences in the number of women per 100 men in China and South Asia. In China the information is shown by province and in India by state. The Indian data are for 1991–93; the Chinese are mostly from the 1997 sample census report or, where these are not available, from the 1990 census.

○ The countries shown here contain 42 per cent of the world's population. They all have a male majority and this explains why, in spite of the female majority elsewhere, women are a minority in the world.

The map shows that in all the provinces of China and in all the states of India except Kerala men predominate. But the level of male majority differs considerably: it is most marked in the northern states of India (especially Punjab, Haryana and Uttar Pradesh), in Pakistan and in the central provinces of China. To explain the male majority in the world, therefore, we need to explain it in these countries, provinces and states.

The explanation includes discriminatory treatment of women from conception to old age. First, the universal male majority of births is accentuated by the growing practice of selective abortions, made possible by the spread of echography. In some areas of China there are now over 20 per cent more boys than girls under 4. Second, there are many reports of selective infanticide against girls. Third, young girls may suffer physical discrimination which affects their survival – in nutrition, access to medical treatment (›34), in the work they are obliged to do and through physical violence. All these forms of discrimination and others can persist throughout a woman's life. So in fact to explain these discriminatory practices is to explain the world's male majority.

■ UNFPA 1997; India Profile 2000; China Dimensions 2000; UNESCAP 2000.

Ratio of female to male child mortality (early 1990s)

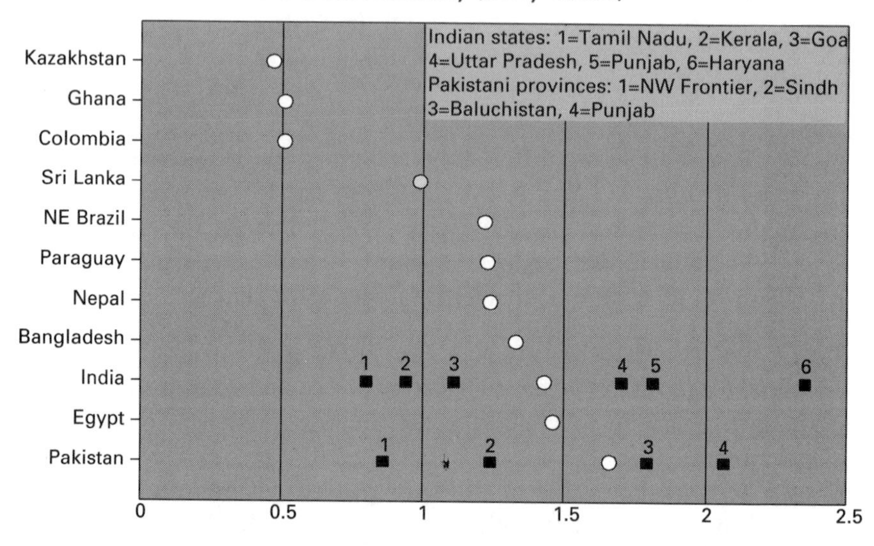

Ratio of female to male children not treated for fever or ARI (early '90s)

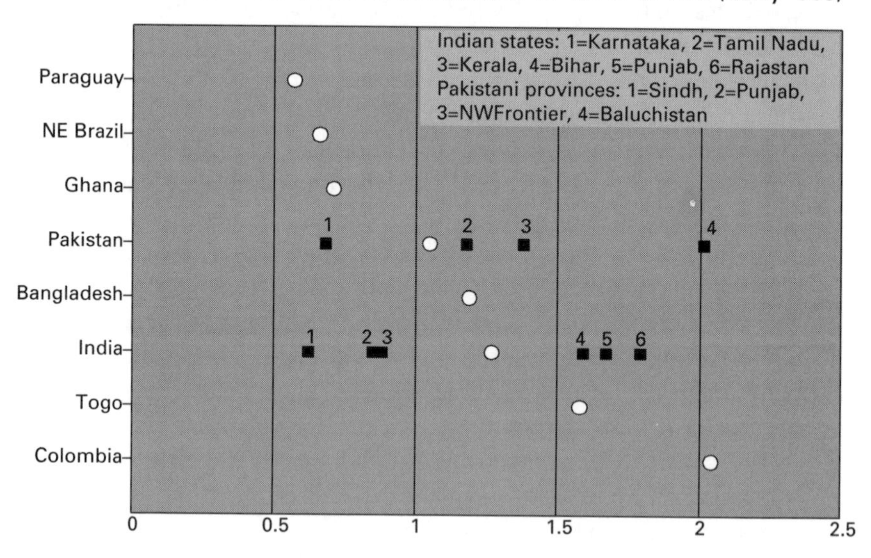

Excess female mortality and some causes

● The top chart shows the ratio between girls' and boys' mortality in selected countries (circles) and in selected states of India and provinces of Pakistan (squares). The bottom graph shows the ratio between girls and boys who do not receive treatment for fever or acute respiratory infection (ARI) in the same countries and is again disaggregated for India and Pakistan.

○ These graphs both confirm and in part explain the excess mortality of women in South Asian countries. The top graph shows that in some countries outside South Asia female mortality is below male mortality but that in all the South Asian countries female mortality is higher. But when the data are disaggregated into states and provinces major differences appear. In Punjab in India female child mortality is 80 per cent higher than male child mortality, and in Haryana it is nearly 2.5 times as high. Yet in Tamil Nadu and Kerala female child mortality is lower than male. So, in addition to selective abortions, this difference in child mortality is probably an important cause of the male majority in the population in these countries.

The lower chart goes some way to explaining the difference in mortality. It shows clearly that girls are discriminated against in the receipt of medical treatment. The usual regional differences are visible: in the Punjab boys have 75 per cent more chance of receiving treatment when they are ill than girls; but in three states, including Kerala, the chances are reversed. The latter states are those with the smallest male majority or even a female majority in the population.

The graph also shows that there are countries in other regions (Togo and Colombia, for instance) where the health discrimination against girls if even worse than in South Asia.

■ Filmer et al. 1998.

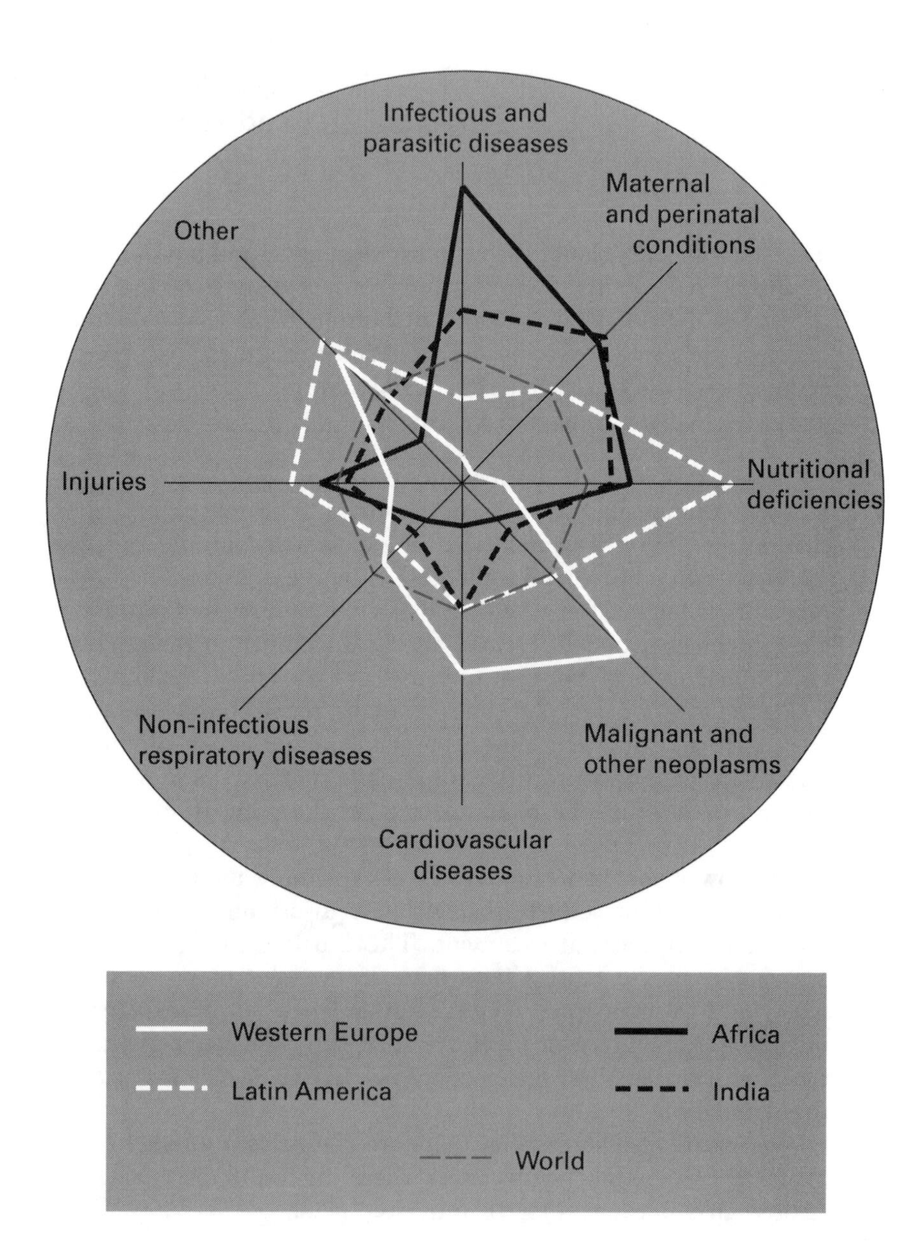

Infectious and parasitic diseases
Maternal and perinatal conditions
Nutritional deficiencies
Malignant and other neoplasms
Cardiovascular diseases
Non-infectious respiratory diseases
Injuries
Other
Western Europe
Latin America
Africa
India
World

● This radial or radar graph allows a comparison of divergences from the world average in the relative importance of eight causes of death. On each axis the prevalence of that cause of death in the world as a whole is indicated by a point at a fixed distance from the origin (the centre). When joined up these points form a regular octagon. The relative importance of the eight causes for each of four regions is marked in the same way on each axis as a multiple of the world level. The irregular octagon formed by joining the points for each region allows easy comparison with the world average.

○ The differences we have seen in life expectancy and the age and sex composition of populations in different parts of the world are very closely related to the pattern of disease. For each of the regions shown in this chart its polygon shows its pattern of fatal disease and disability in comparison with the world average. So the polygon for Africa is very sharply biased towards infectious and parasitic disease and conditions of mother and child related to birth. Nutritional deficiencies and injuries are also a more important cause of death than in the world as a whole. Western Europe by contrast has a low relative prevalence of infectious, birth-related and nutrition-related causes of death but a correspondingly very high relative importance of heart disease and of cancer.

A common misinterpretation of the information in this chart is that cardiovascular disease and cancer are caused by development. In fact an adult in Africa has a greater probability of dying from these causes than an adult in Western Europe. These look like diseases of development because developed countries have a greater life expectancy and so people die of disorders which typically appear later in life.

■ WHO 1999.

Tuberculosis

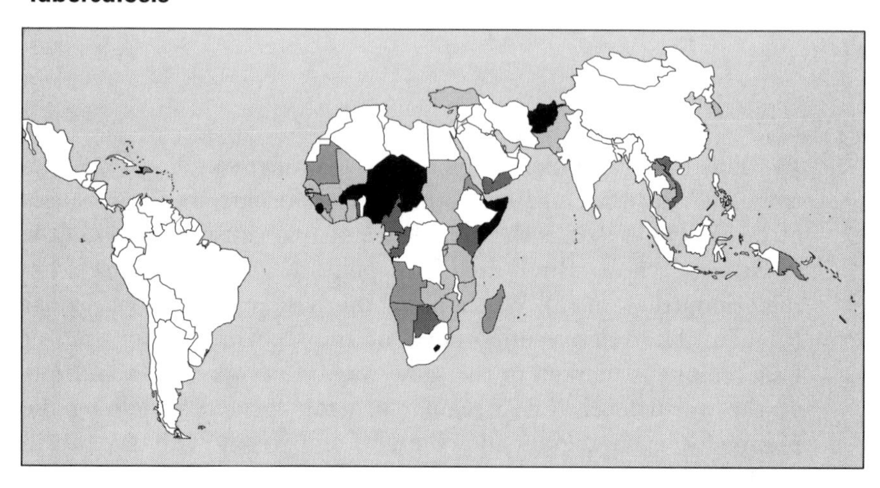

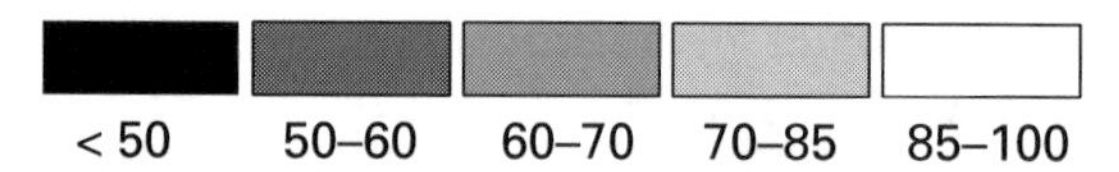

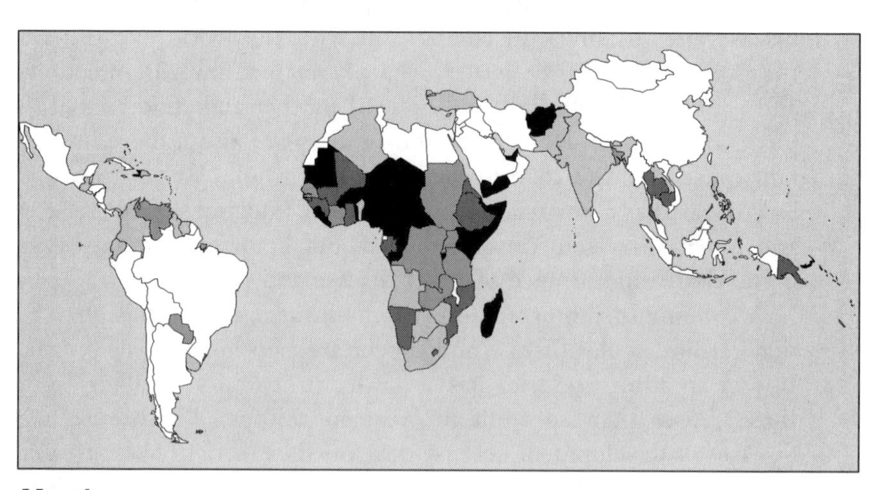

Measles

● The maps show the extent of inoculation of 1-year-old children against tuberculosis and measles in the period 1995–97.

○ Two of the major diseases that cause a very high relative level of deaths from infectious diseases in poor countries are tuberculosis and measles. Like many other infectious diseases they are to a great extent preventible by inoculation. And inoculation programmes, financed by both national governments and international agencies, have in many parts of the world had an extraordinary success and have been implemented in extremely difficult conditions. Here it can be seen that most of Latin America and large parts of Asia have achieved more than 70 per cent coverage of children in inoculation programmes. Countries with less than 50 per cent coverage are largely in Africa where a combination of war, poverty and the absence of public health infrastructure have sometimes made it difficult to carry out mass inoculation programmes.

International inoculation programmes were successful in eliminating smallpox and, although delayed, there is hope that the World Health Organization's plan to eliminate polio will be successful before the year 2005.

Mass inoculation is a spectacular and effective measure. But it can only go a limited way in curbing child deaths. At least as much depends on general public health measures such as improved sanitation and drinking water, which will eliminate infectious ailments for which no remedy is available via inoculation.

■ UNDP 1999.

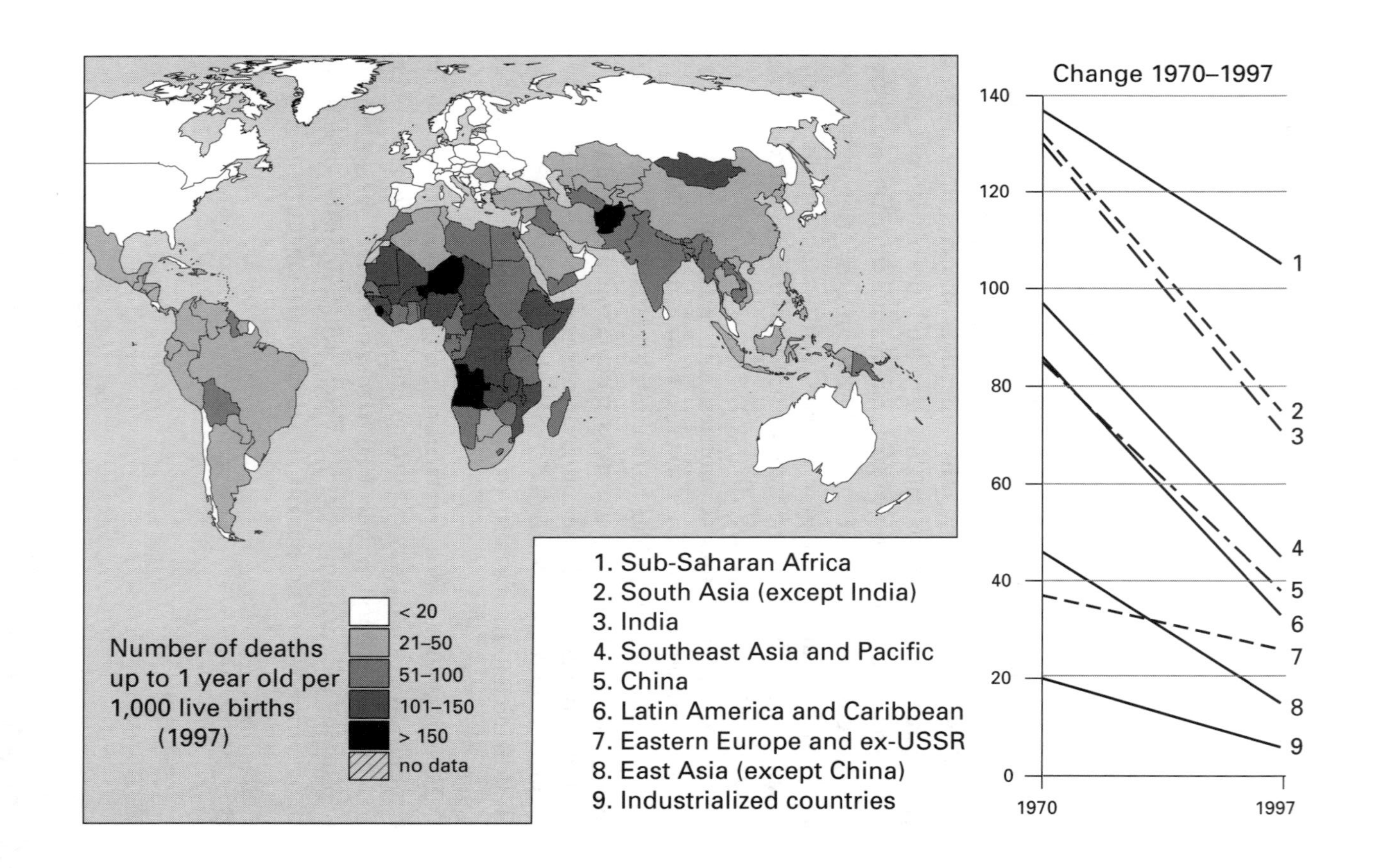

Number of deaths up to 1 year old per 1,000 live births (1997)
< 20
21–50
51–100
101–150
> 150
no data
1. Sub-Saharan Africa
2. South Asia (except India)
3. India
4. Southeast Asia and Pacific
5. China
6. Latin America and Caribbean
7. Eastern Europe and ex-USSR
8. East Asia (except China)
9. Industrialized countries
Change 1970–1997
140
120
100
80
60
40
20
0
1970
1997
1
2
3
4
5
6
7
8
9

● The map shows the infant mortality rate (number of deaths up to 1 year old per 1,000 live births) by country for the year 1997. The graph shows the change in this figure between 1970 and 1997 for the nine mentioned groups of countries.

○ The figures for infant and child mortality may appear to be encouraging ones. In every area of the world there has been a marked reduction in infant mortality during the last thirty years. The causes have been general improvements in public health and hygiene and specific programmes such as mass inoculation (described in the previous chart). The fact that children have a lower probability of dying today is an important piece of evidence that some development has been happening even in the worst-performing regions. East Asia has matched its very rapid industrial and economic growth with equally rapid improvement in indicators such as this one. A child in East Asia now has very little more chance of dying than in a European country or in the USA. And a few countries, most notably Cuba, have shown that it is not necessary even to have economic growth to reduce child mortality to minimum levels. Strong popular commitment to public health programmes and well-targeted spending is enough.

But there is also a disturbing aspect to these figures. Despite large absolute improvements, the rate of infant mortality today in the poorer countries is worse relatively to the richer countries than it was thirty years ago. In 1970 sub-Saharan Africa had 7 times the level of infant mortality as the industrialized countries; now it has about 17 times the level. Within a framework of general improvement the level of international inequality has actually grown considerably.

■ UNDP 1999.

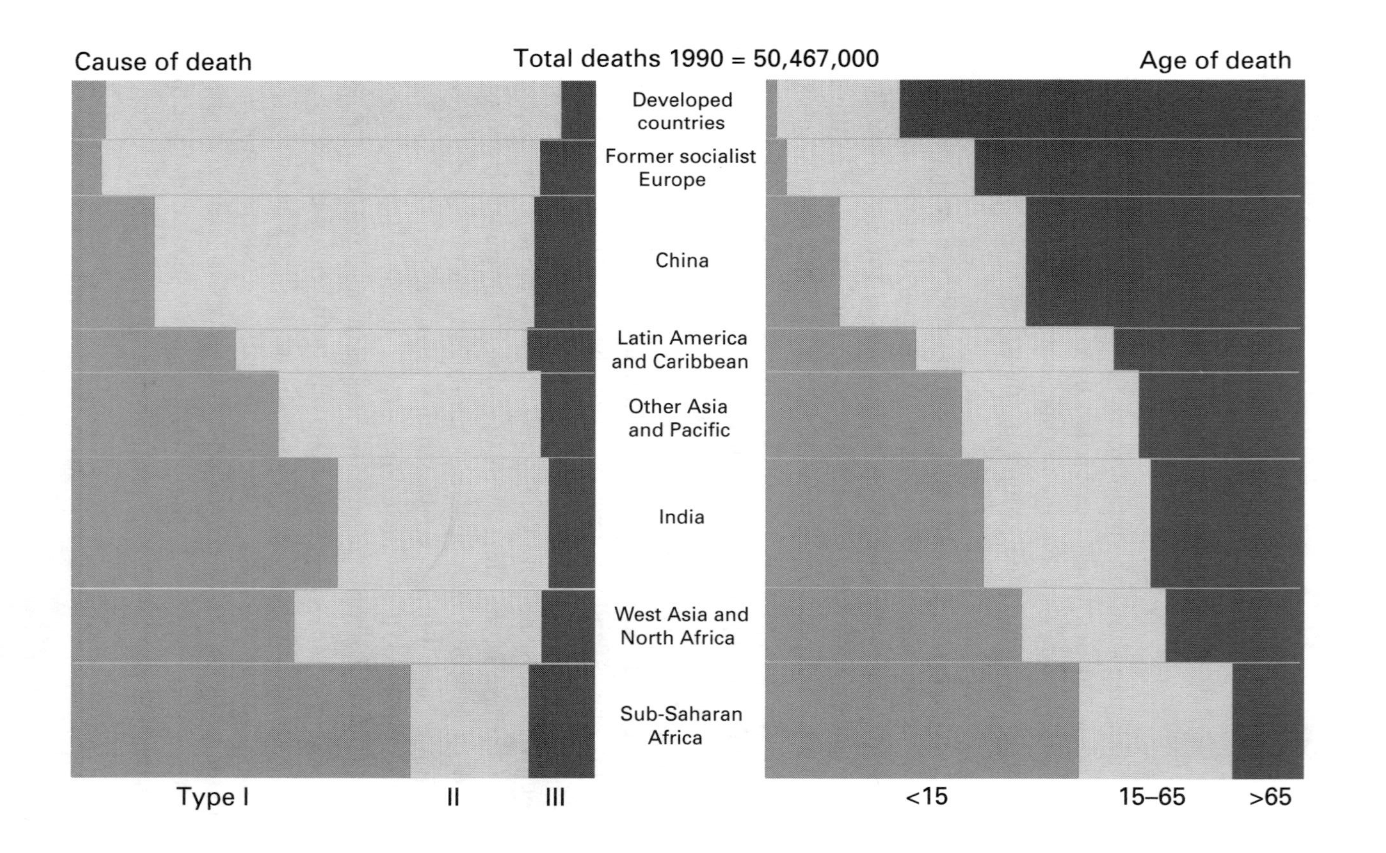

Cause of death
Total deaths 1990 = 50,467,000
Age of death
Developed countries
Former socialist Europe
China
Latin America and Caribbean
Other Asia and Pacific
India
West Asia and North Africa
Sub-Saharan Africa
Type I
II
III
<15
15–65
>65

● Both the shaded areas represent the total number of deaths that occurred in the world in 1990. They are both divided in horizontal bands according to the number of deaths in each of eight regions. The left-hand chart divides them vertically by three types of cause of death: I, infectious diseases and causes connected with birth and maternity; II, non-infectious medical conditions, most notably cancers and heart disease; III, accidents, killing and suicide. The right-had chart divides them by the age of those who died.

○ There is a strong relationship between level of development and the pattern of causes of death. In the poorest regions the majority of deaths have type I causes. In the most developed regions these causes have been virtually eliminated and the enormous majority of deaths have type II causes. Type III causes produce a relatively small percentage of deaths, varying considerably between regions but without any apparent relationship to the level of development. The difference highlighted by these figures can be seen most clearly by comparing sub-Saharan Africa, where 65 per cent of deaths have type I causes, with the OECD countries where only 5 per cent have type I causes. There is a fairly clear relationship between the percentage of deaths at ages below 15 and the percentage of type I causes.

Changes in the relative importance of the three types of cause of death are also related to the sex composition of deaths. Up to the age of 45 deaths with type I causes are predominantly female. Type II and III causes affect men disproportionately at all ages, with the already mentioned very high peak in the male/female ratio of type III causes up to the age of 40 (‹ 31).

■ Murray and Lopez 1996.

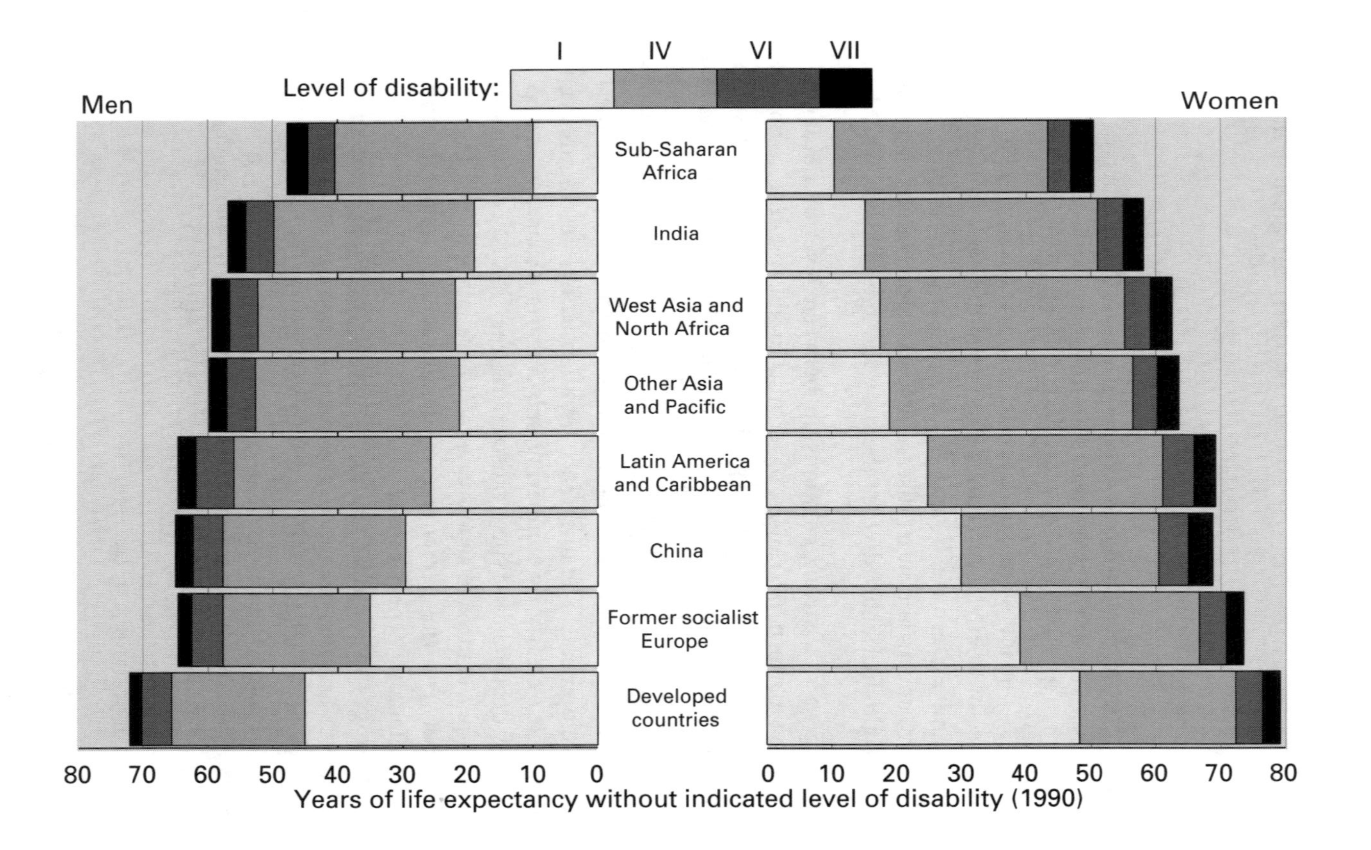
Level of disability:
I
IV
VI
VII
Men
Women
Sub-Saharan Africa
India
West Asia and North Africa
Other Asia and Pacific
Latin America and Caribbean
China
Former socialist Europe
Developed countries
80 70 60 50 40 30 20 10 0
0 10 20 30 40 50 60 70 80
Years of life expectancy without indicated level of disability (1990)

● The graph summarizes the methodology that produces the estimates of disability-adjusted life expectancy in graph 23. The length of the bars represent life expectancy for men (left) and women (right) in 1990. The bars are divided to indicate how expected life is divided into years without and years with different levels of disability, defined after an exhaustive discussion within a panel of non-professional people about how severely they rated different conditions. They divided disabilities into seven degrees of severity: for example, vitiligo on the face is considered a level I disability, deafness is level IV, while quadriplegia is regarded as level VII.

○ Years of life are not, of course, homogeneous and so life expectancy alone is not an ideal measure of health or physical welfare. We could measure welfare better if we also had indicators of morbidity or disability. The international study *The Global Burden of Disease* (*GBD*) has tried to produce the first usable worldwide inventory of disability, combining it with their mortality estimates in order to produce a global measure of the burden of disease. While far from perfect, this estimate can be used to show the way beyond life expectancy to more discriminating indicators of human development.

For women life expectancy in general ranges from 80 years in the OECD countries to 45 years in sub-Saharan Africa (for men between 73 and 48 years). But the *GBD* estimates that the equivalent range of expectancy of life without any level of disability ranges from 48 years to only 10 years (for men 45 to 10 years). This means that the average citizen of the poorest countries can, according to these estimates, expect to live only about 10 years free of some significant level of disability.

■ Murray and Lopez 1996.

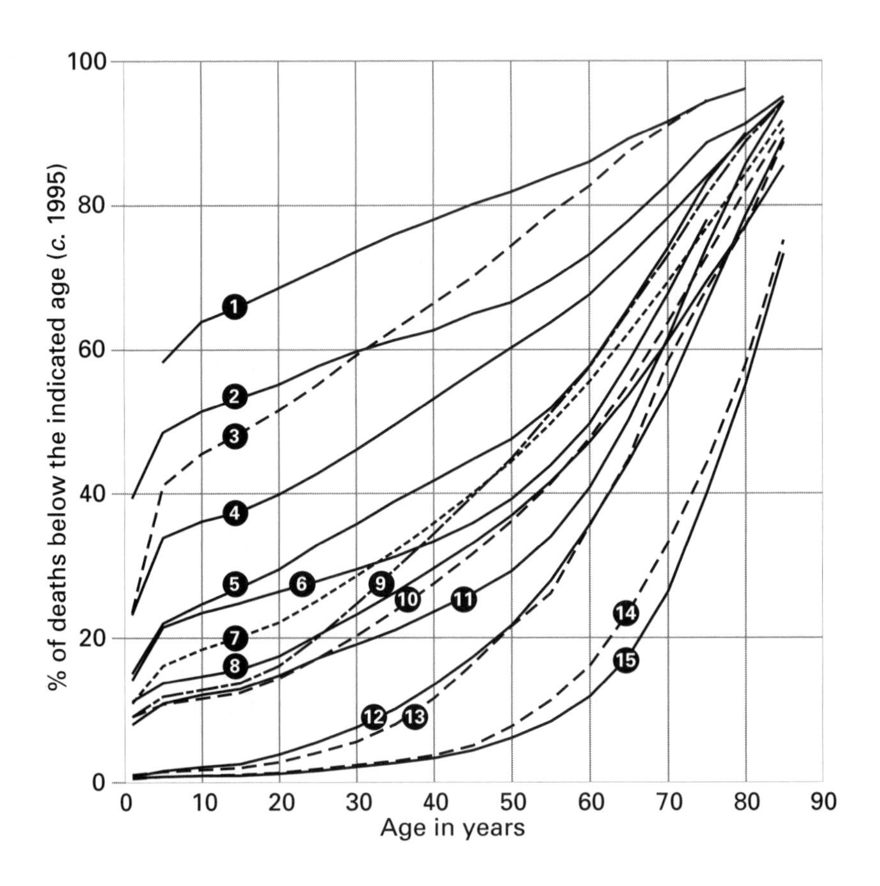

1. Mali
2. Pakistan
3. Central African Republic
4. Guatemala
5. Morocco
6. Egypt
7. Philippines
8. Mexico
9. South Africa
10. Brazil
11. China
12. South Korea
13. Russian Federation
14. Japan
15. UK

● The curves on the graph show for the fifteen indicated countries the proportion of the people who die (shown on the vertical axis) who are below the age indicated on the horizontal axis.

○ The age pattern of mortality is that very young children die in relatively significant numbers but then mortality falls considerably for people between 5 and 35 years old; it then rises again to reach extremely high levels for the older age groups. That universal pattern suggests that, although some countries have tragically high infant mortality, death is still basically what happens to people who have grown old. If, however, we look not at mortality rates but at the deaths that actually take place, a different picture emerges. In a country with high infant mortality rates and a high proportion of children in the population, death is not predominantly something suffered by old people but by children. Though many of the poorest countries lack the appropriate figures, some startling examples exist: in Mali (curve 1) almost 60 per cent of the people who die are under 5 years old; in Pakistan (curve 2) 40 per cent of those who die are under 1 and 50 per cent under 5. In the UK, by contrast (curve 15), almost three-quarters of those who die are over 70 years old and almost half are over 80. This conveys the enormous difference that the phenomenon of death assumes in different countries. In Britain, like other rich countries, it is exceptional for a young person to die. In many parts of Africa the death of an old person is the exceptional event. When, as a matter of course, children die before their parents, the presence of death must be especially oppressive.

■ UN 1999.

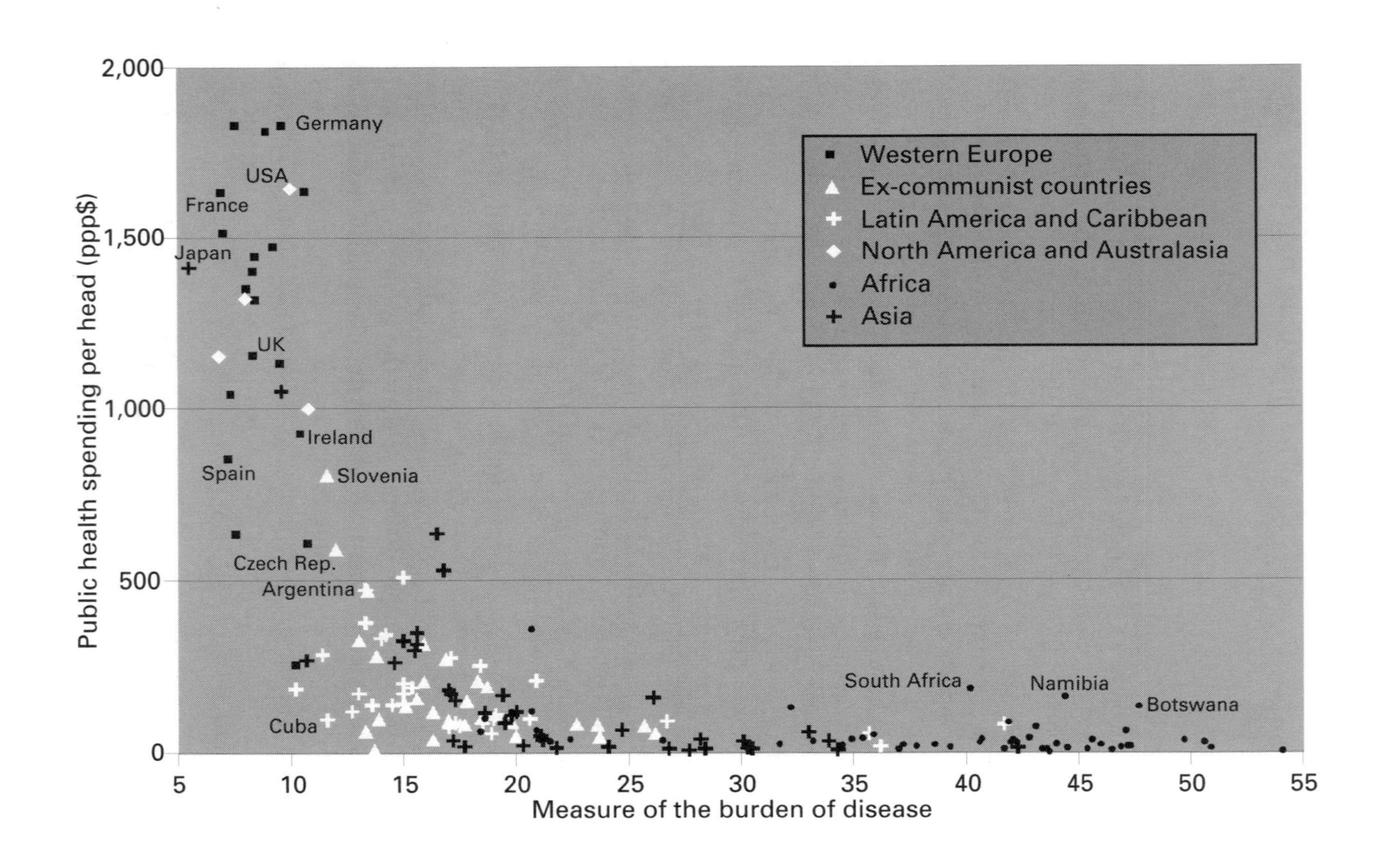

Public health spending per head (ppp$)
2,000
1,500
1,000
500
0
Measure of the burden of disease
5
10
15
20
25
30
35
40
45
50
55
Western Europe
Ex-communist countries
Latin America and Caribbean
North America and Australasia
Africa
Asia
Germany
USA
France
Japan
UK
Ireland
Spain
Slovenia
Czech Rep.
Argentina
Cuba
South Africa
Namibia
Botswana

● Each point represents one country and shows both the amount its government spent per head on health in 1998 (on the vertical axis) and a measure of the burden of disease, or health need (on the horizontal axis). This measure is calculated by taking the difference between the disability-adjusted life expectancy of the country and 80 (‹23). It assumes that the best situation would be for everyone to enjoy eighty years of life without disability due to illness. The higher the figure, therefore, the greater the difference from this ideal, and so the greater the burden of disease. The points are distinguished according to region and some individual countries are identified.

○ It is easily seen from this graph that there is a very strong relationship between health spending and health need between the countries of the world: the greater the health need, the less the health spending. The WHO calculates that the countries of the South suffer as much as 92 per cent of the total burden of disease in the world (measured by lost **DALYs**) while about 80 per cent of health spending takes place in the North. This graph details this relationship.

One interpretation of it is as an indictment of the complete rupture between need and spending, so characteristic of the present world economic system. Another interpretation is more optimistic: more health spending is connected to less burden of disease and so it is an argument to give more priority to health. A third interpretation is even more optimistic: there is a wide range in the spending per head in the countries with low burdens of disease. This suggests that some have found the secret of producing good results but spending less. Almost certainly this secret of wise spending includes giving emphasis to egalitarianism in the health system.

■ WHO 2000.

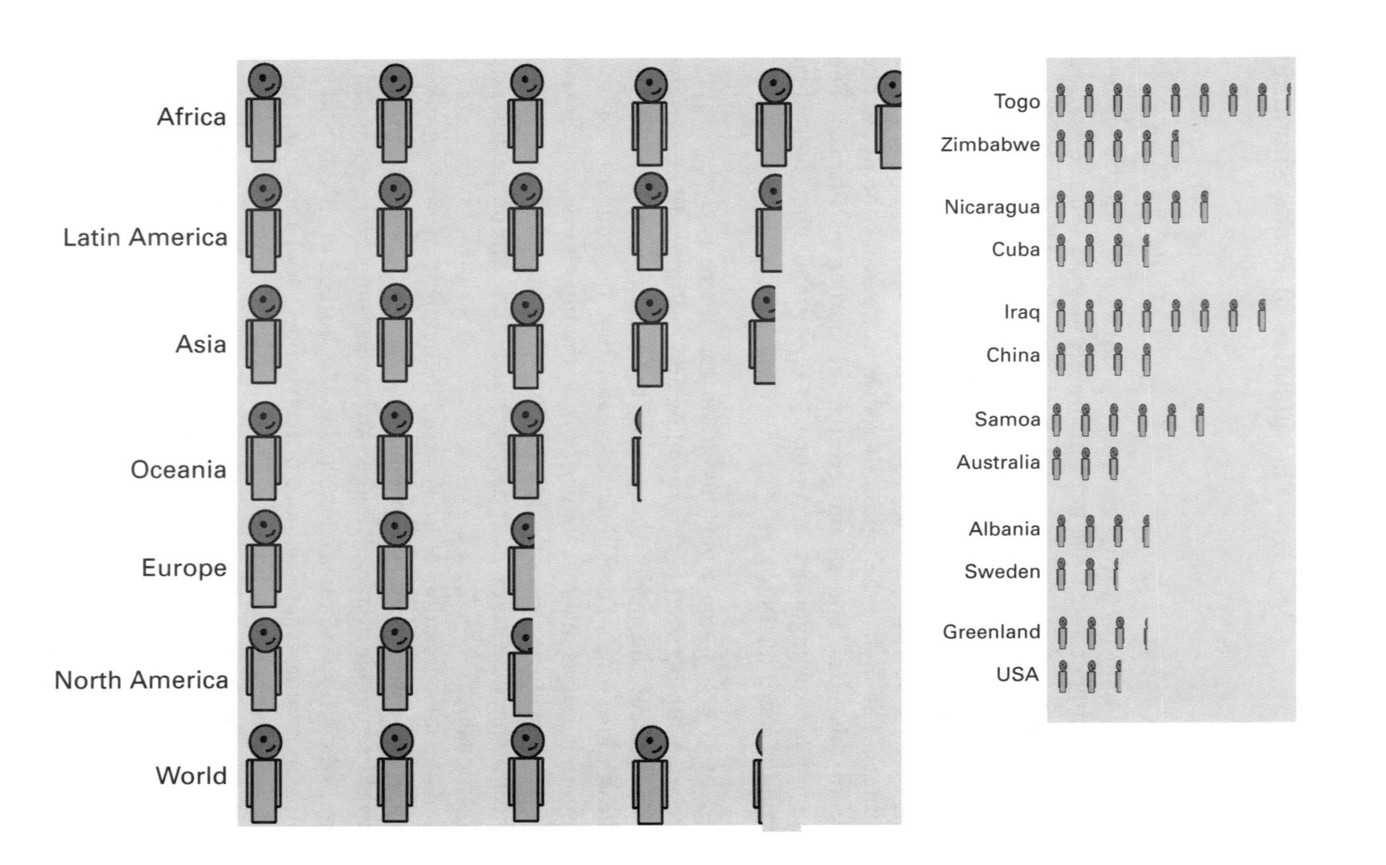
Africa
Latin America
Asia
Oceania
Europe
North America
World
Togo
Zimbabwe
Nicaragua
Cuba
Iraq
China
Samoa
Australia
Albania
Sweden
Greenland
USA

The size of families and households

● The left-hand chart shows the average size of household in the regions mentioned; and the smaller graph at the right shows for each region the figures for the countries with the largest and the smallest household size.

○ These data show that there is a major difference in the size of families and households between countries, generally assumed to be systematically related to development. Research on the history of the family, however, has cast major doubt on the existence of a clear pattern. Nonetheless families are often larger in poor countries partly because of the higher birth rate. And they have diminished in size in richer countries partly as a result of declining numbers of children per parent. But the size of household has diminished for other reasons as well. In particular, generations live together less and less. People in richer countries leave home to lead independent lives at a younger age and only exceptionally at a later date invite their parents to become part of their households. Elderly people in the more developed countries tend to live with their partners, alone or in institutions. This is possible partly because of the introduction of universal old-age pensions. In Britain at present only 8 per cent of people over 65 years of age live with their children, while in South Korea 77 per cent still do so. Solidarity and the sharing of labour within households may in many circumstances aid the collective survival of all its members. But also some of the worst examples of inequality and exploitation take place within households and they tend to be the examples which for very obvious reasons are least investigated. Oppressive inequalities can exist in families regardless of their size or number of generations.

■ UN Centre for Human Settlements 1996.

Land, agriculture, food and hunger IV

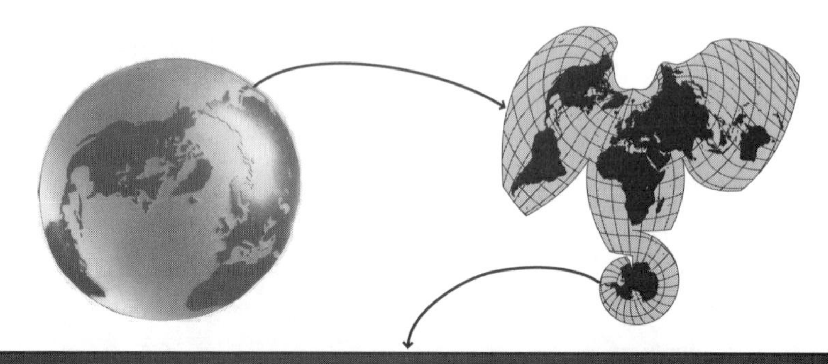

surface area = 510,073,000 km^2

water = 361,132,000 km^2

land = 148,941,000 km^2

without vegetation = 59,098,000 km^2

with vegetation = 89,843,000 km^2

forest = 41,690,000 km^2

permanent pasture = 33,645,000 km^2

cultivated land = 14,508,000 km^2

● The equal area projection of the earth at the top right is by Eric Dudley of MapInfo (see **The data, the graphs and the text**). The squares represent a redivision of this area into the categories indicated.

○ These are some basic facts about our planet which determine part of the objective physical conditions of survival and affect the inequalities between humans. Some 70 per cent of the surface of the planet is water, which is employed in numerous useful and necessary ways. As a direct source of food, the primary subject of this section of the book, water is much less important than land. It supplies about 6 per cent of the proteins consumed and about 2 per cent of the calories.

Fresh water also, of course, performs necessary functions in maintaining the ability of the land to produce food and in the maintenance of human health and welfare.

A little food and many other positive things come from the forests. But well over 90 per cent of human food supply comes, and will continue to come, from pasture and cultivated land, which together make up less than 10 per cent of the earth's surface.

The proportions of the different categories in this chart are not fixed. Each of the frontiers is a line of conflict and of advance or retreat, due either to natural forces or to the effects of human activities. Global warming will reduce the area of land; deforestation will reduce the area with vegetation. And there is a permanent social struggle over the use of land at the many frontiers between forest, pasture and cultivated land.

■ WRI 1998.

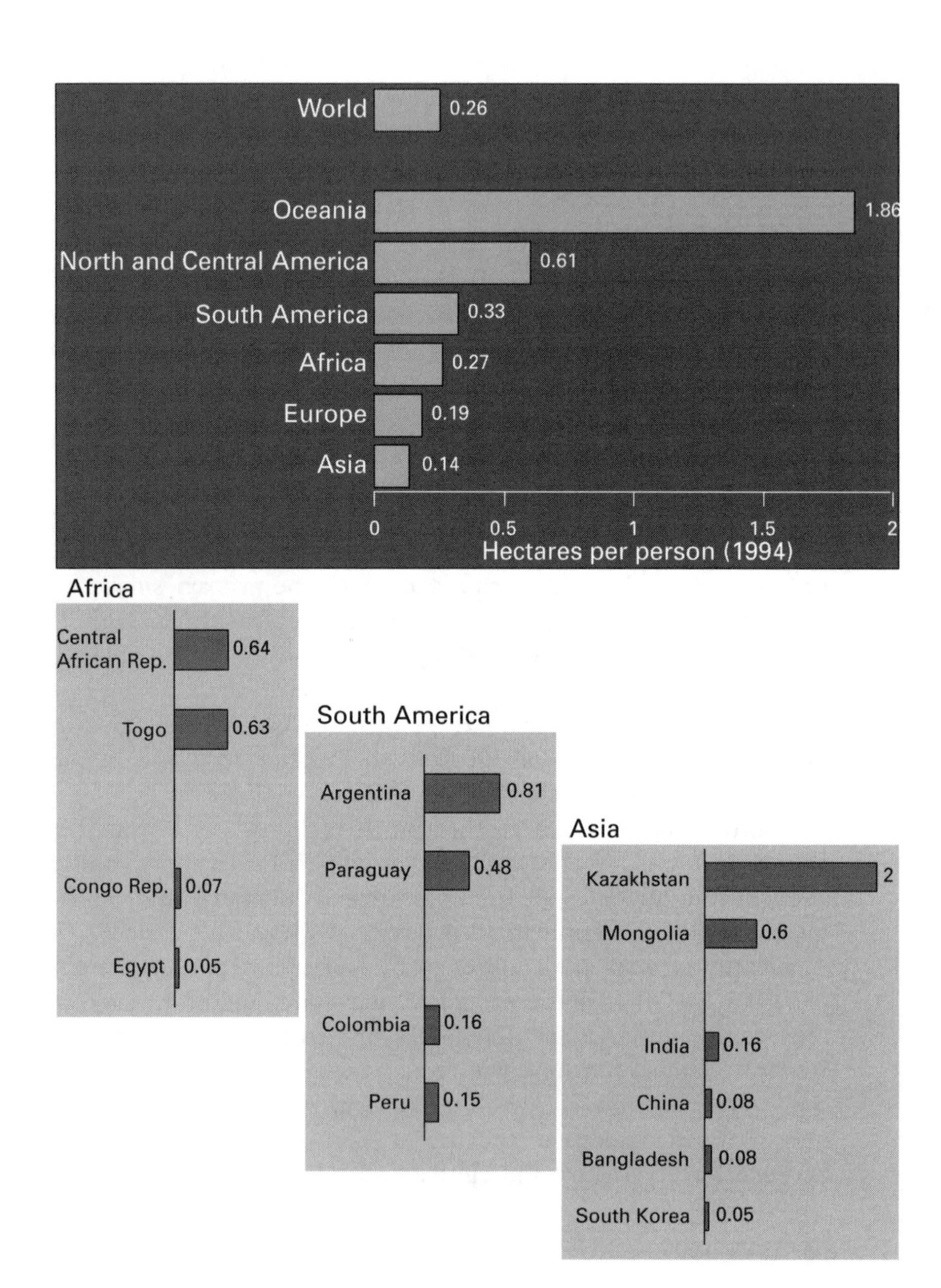

World 0.26
Oceania 1.86
North and Central America 0.61
South America 0.33
Africa 0.27
Europe 0.19
Asia 0.14
0
0.5
1
1.5
2
Hectares per person (1994)
Africa
Central African Rep. 0.64
Togo 0.63
Congo Rep. 0.07
Egypt 0.05
South America
Argentina 0.81
Paraguay 0.48
Colombia 0.16
Peru 0.15
Asia
Kazakhstan 2
Mongolia 0.6
India 0.16
China 0.08
Bangladesh 0.08
South Korea 0.05

● The upper chart shows cultivable land per person by continent in hectares. The lower charts show the same in the two countries with the largest and the two with the smallest amounts in three of the continents.

○ In the world as a whole about a quarter of a hectare of cultivable land is available per person. But it is geographically distributed in a very different way from the population. So Oceania has about ten times the availability of Europe and over ten times that of Asia. The lower charts show also that the continents are not homogeneous: Kazakhstan has over 20 times as much land per head as Bangladesh, the Central African Republic over 12 times as much as Egypt.

In modern times surprisingly little connection has existed between the availability of land and national wealth. The USA and Australia have developed on the basis of agriculture but others (such as Britain and Japan) have exported their industrial products and imported food.

From its small allocation of cultivable land an expanding human society has to produce nearly all its food. But land is only one of the factors needed to produce food. So when land is scarce it is used more intensively, and that means using it with greater quantities of the other factors: labour, water, fertilizer, pesticides and machinery.

■ WRI 1998.

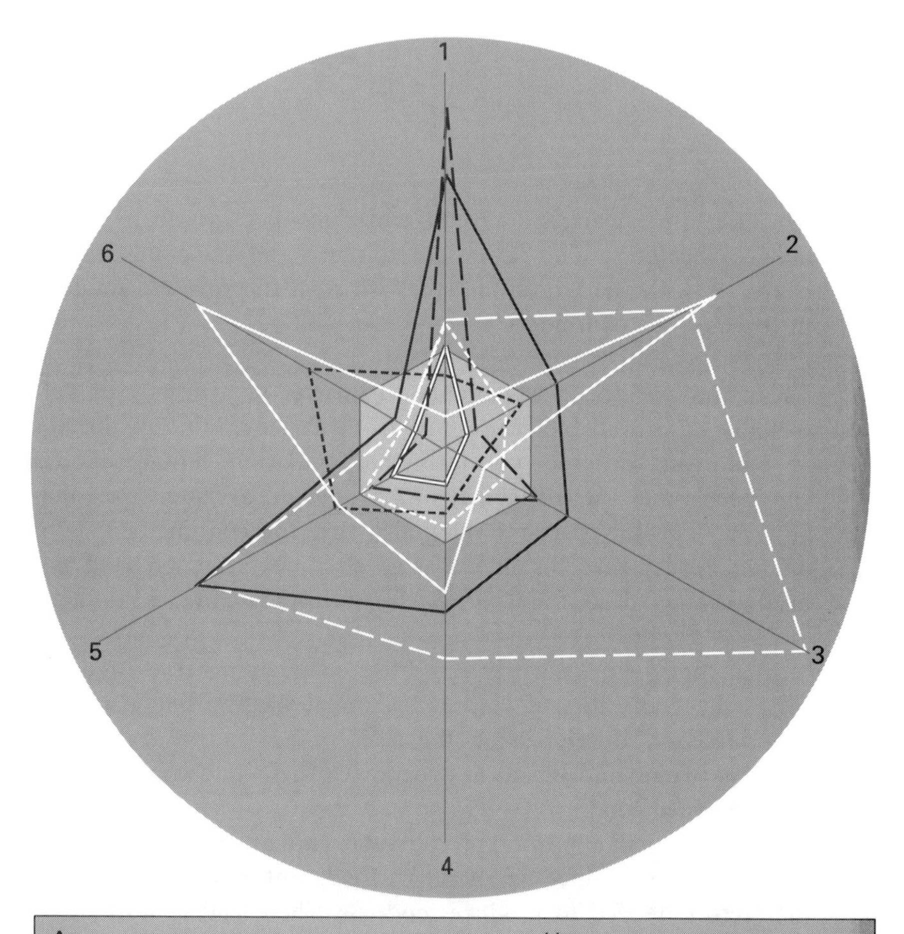

Axes

1. Hectares of land per head
2. Kg of fertilizer per hectare
3. Tractors per hectare
4. Cereal yield (kg/ha)
5. Roots and tubers yield (kg/ha)
6. Percentage of land irrigated
(kg/ha = kilograms per hectare)

Hexagons

USA
Russia
France
India
China
South America
Africa

● This is another radial or radar graph like number 36. In this case the six axes represent some aspect of the inputs or outputs of agriculture in the seven countries or regions described, each of which has its own hexagon. The world average for each variable is set at the same fixed distance along the axis, producing a regular hexagon. The others show not absolute differences but the comparative level of each variable in relation to the world average.

○ This chart shows up the great contrasts that exist in the agricultures of the world. Africa, with about the world average of available land, has an agriculture which is less productive than the world average but at the same time uses less inputs. In China a relatively high yield is combined with a high use of irrigation and fertilizers. In France very high yields are obtained with the use of large quantities of fertilizer and machinery. The USA obtains high yields with little irrigation but a fairly high use of tractors and fertilizer. In Russia agriculture is extensive and low yielding, but also with few manufactured inputs aside from tractors. India, like China, has a large quantity of irrigated land but used far less fertilizer and has lower yields. South America is closest to the world average, although it is slightly above in availability of land and considerably below in the proportion of irrigated land.

The graph is a good illustration of how different inputs can be substituted for each other and of how really diverse an activity is agriculture in the world.

■ WRI 1998.

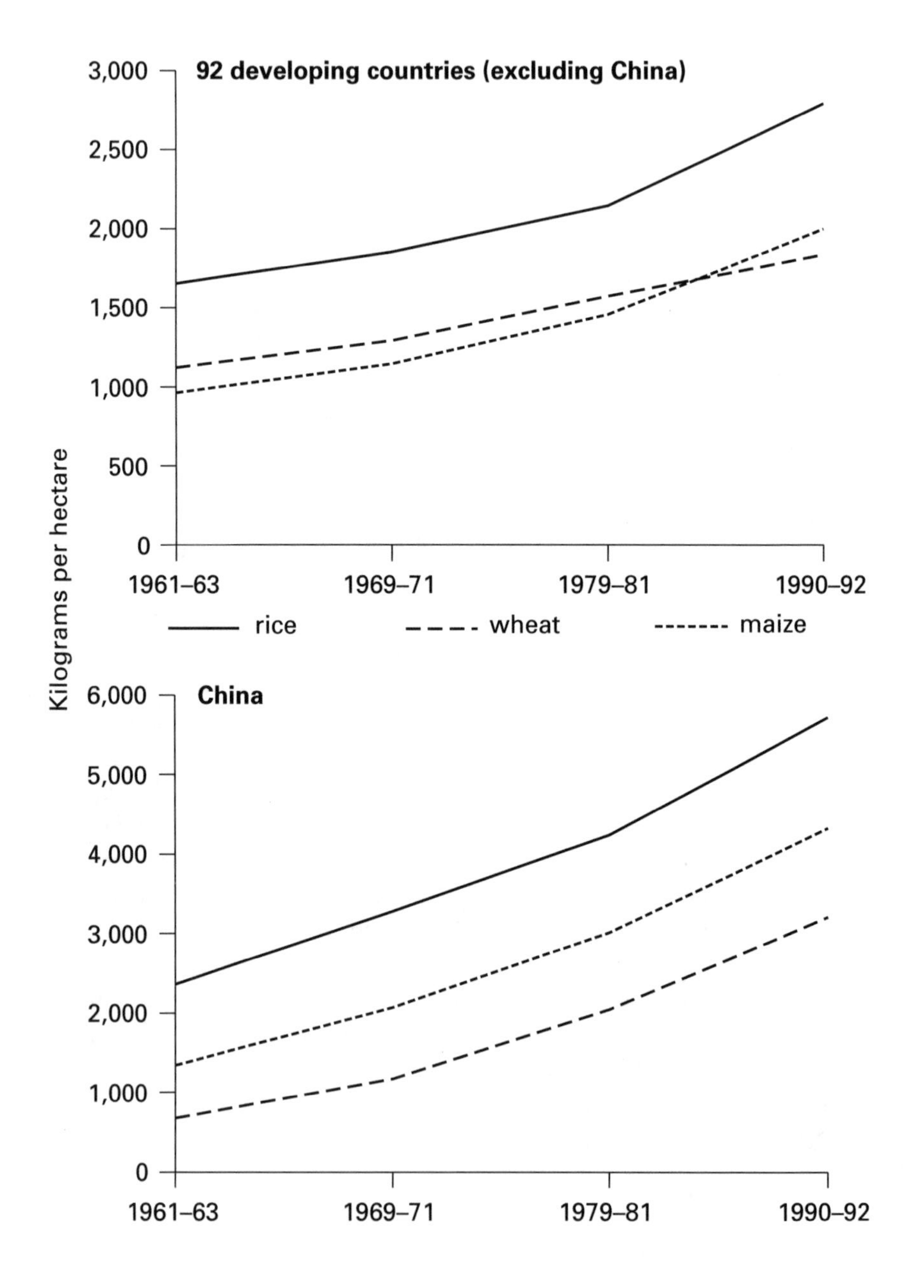
92 developing countries (excluding China)
3,000
2,500
2,000
1,500
1,000
500
0
1961–63
1969–71
1979–81
1990–92
rice
wheat
maize
Kilograms per hectare
China
6,000
5,000
4,000
3,000
2,000
1,000
0
1961–63
1969–71
1979–81
1990–92

Crop yields and the green revolution

● The graphs show the evolution since the early 1960s of cereal yields in agriculture in the developing countries. The upper chart shows the average for ninety-two countries and the lower one China alone. The figures are averages for the years mentioned.

○ During the thirty years shown in these graphs average yield per unit of land in cereal production in the developing countries has risen by about 50 per cent. This, along with extensions in the land area where it is still possible, has led to a large increase in food production in aggregate, partly due to what has been called the 'green revolution'. This means the use of new varieties of seeds for cereals, developed in research institutes and by grain companies during the 1950s and 1960s. These varieties produced higher yields per hectare than existing varieties as long as they were treated in certain ways. They required more water, fertilizer and pesticides. So for farmers the result may be more rewarding but it is also initially more costly. Many people have argued that this has accelerated inequality since richer farmers have been able to use the new varieties more readily than poorer farmers, and as a result land holdings have become more concentrated and in some places landlessness has grown. There is much disagreement, however, in India about the effect of the green revolution on inequality.

The green revolution was not a one-off event. Major new varieties with even higher potential yields, which are more tolerant of pesticides and which require less irrigation, have now appeared and farmers are also being offered genetically modified seeds. Critics of the process argue that it overencourages monocultures, destroys biological diversity and leads to greater dependence of farmers on seed companies.

■ WRI 1998; FAO 1997.

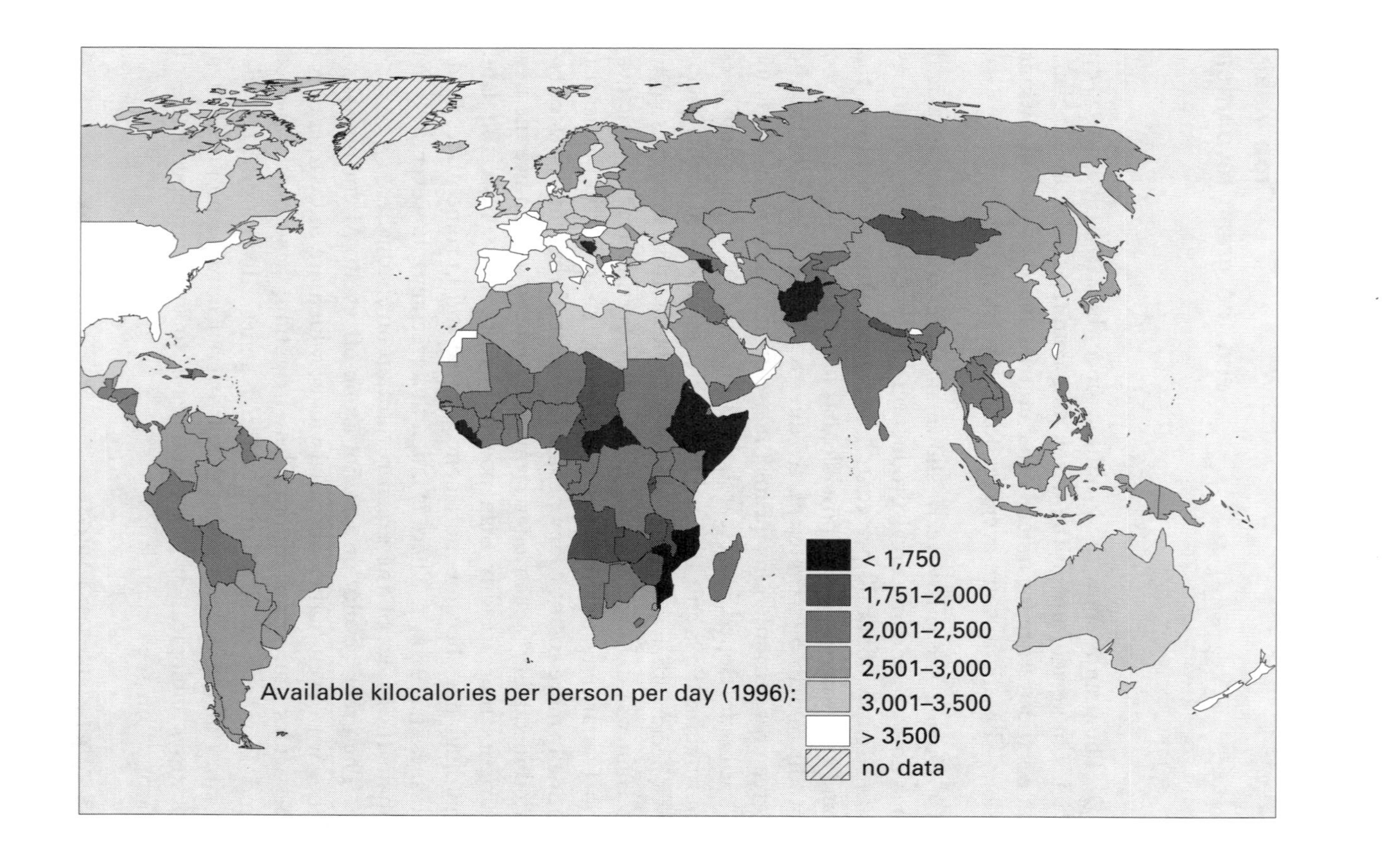
Available kilocalories per person per day (1996):
< 1,750
1,751–2,000
2,001–2,500
2,501–3,000
3,001–3,500
> 3,500
no data

Unequal access to food: calorie supply

● The map shows levels of kilocalorie availability per head for the year 1996. The figures are produced annually by the FAO and they represent the total registered or estimated national production of food, plus food imports, converted into calories, reduced by an arbitrary factor to account for waste and divided by the population.

○ The figures on which this map is based provide an extremely rudimentary idea of international differences in one aspect of nutrition – food energy supply, or quantity of calories. It is not possible to reduce calorie requirements to a simple number per person since it depends on each person's metabolism, physical activity level, age and weight. But as a very rough approximation, if these figures are at all accurate, they suggest that in countries with an average daily calorie supply of less than 2,000, whatever the distribution of the available food, there is almost certain to be a widespread deficiency in food energy supply. Where the figure exceeds 3,000 a day then, unless there is extreme inequality in distribution, food energy supply should be generally sufficient. Between 2,000 and 3,000 the degree of undernourishment will depend strongly on the degree of inequality in the distribution of food. Since these figures are national averages per head they contain no information about distribution. At best, therefore, they are very rough guides indeed to the adequacy of food intake. Their advantage is that they are published regularly; and the movements from year to year may be more informative than the actual levels.

■ FAO 1997.

a. Number of undernourished people by major region

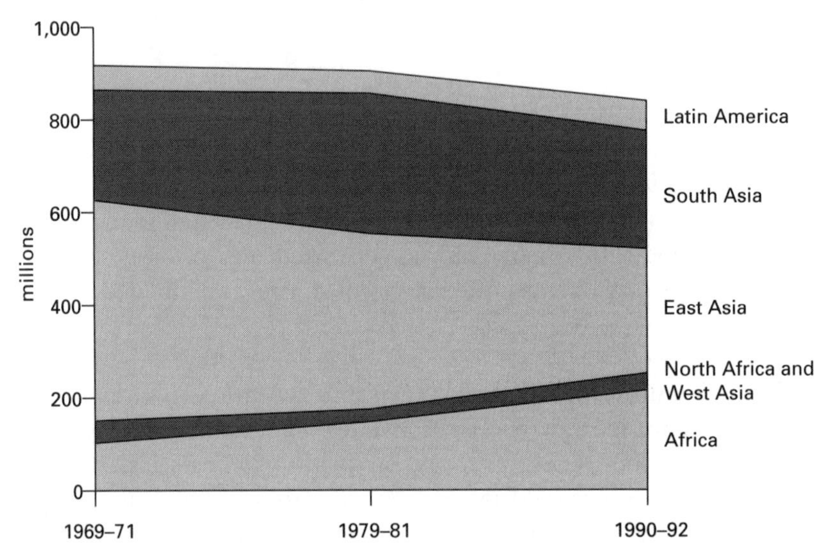

b. Percentage of the population undernourished by major region

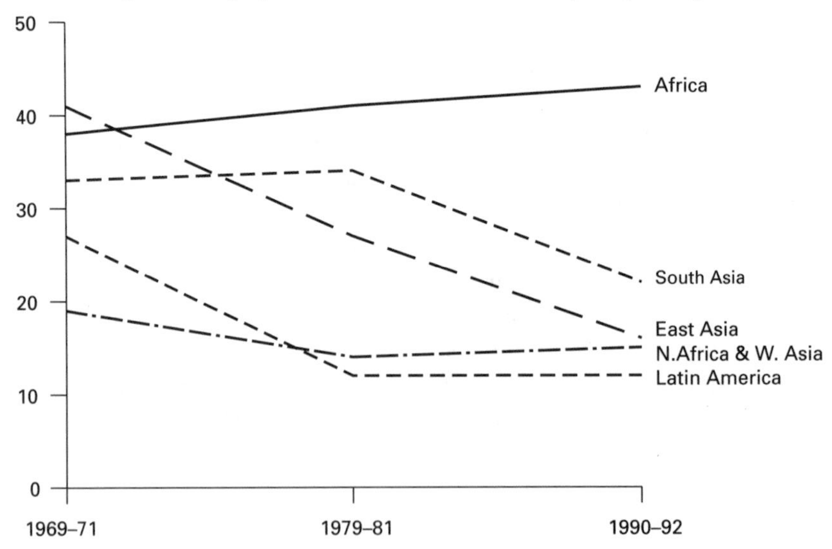

● These two charts summarize the latest information provided by the FAO about the extent of undernourishment (insufficient food energy supply) in the world and its regions. The information is based not on the annual total calorie availability of the previous chart but on periodic World Food Surveys in which more detailed sample surveys of food consumption in households are conducted in different parts of the world and then unified to produce a general picture.

○ The most recently published FAO World Food Survey was conducted between 1990 and 1992. These are the estimates that it produces for the degree of undernourishment in the world. The FAO concludes that a little over 800 million people (about one-sixth of the world's population) suffered from undernourishment in 1990–92, a figure in total a little lower than at the time of the previous estimates one and two decades earlier. The number of undernourished people remained about the same in Latin America, fell in Asia, especially East Asia, and rose in Africa. The percentages of people estimated to be suffering undernourishment are shown in the lower chart. Again the declines in Asia and the increase in Africa are striking. Unlike the national figures in the last graph, these figures do in principle take distribution into account. They have, however, been strongly criticized on the grounds that the surveys are inadequate and that anthropometric data (such as the weight of children, and weight to height ratios) would give a more accurate picture of undernourishment than calorie intake. They remain, however, the most 'official' estimates of hunger that exist.

■ FAO 1996.

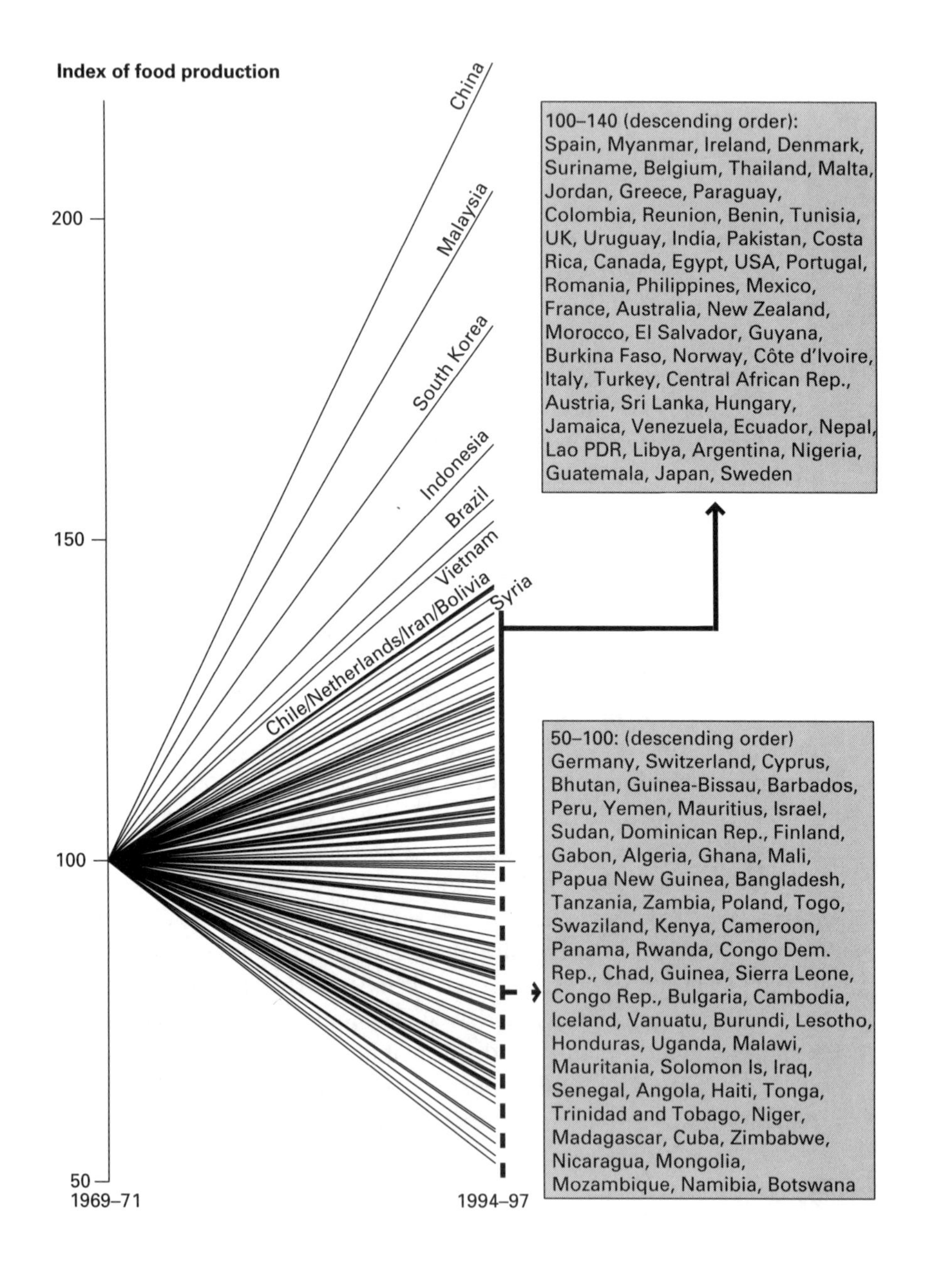

Index of food production
200
150
100
50
1969–71
1994–97
China
Malaysia
South Korea
Indonesia
Brazil
Vietnam
Chile/Netherlands/Iran/Bolivia
Syria
100–140 (descending order):
Spain, Myanmar, Ireland, Denmark, Suriname, Belgium, Thailand, Malta, Jordan, Greece, Paraguay, Colombia, Reunion, Benin, Tunisia, UK, Uruguay, India, Pakistan, Costa Rica, Canada, Egypt, USA, Portugal, Romania, Philippines, Mexico, France, Australia, New Zealand, Morocco, El Salvador, Guyana, Burkina Faso, Norway, Côte d'Ivoire, Italy, Turkey, Central African Rep., Austria, Sri Lanka, Hungary, Jamaica, Venezuela, Ecuador, Nepal, Lao PDR, Libya, Argentina, Nigeria, Guatemala, Japan, Sweden
50–100: (descending order)
Germany, Switzerland, Cyprus, Bhutan, Guinea-Bissau, Barbados, Peru, Yemen, Mauritius, Israel, Sudan, Dominican Rep., Finland, Gabon, Algeria, Ghana, Mali, Papua New Guinea, Bangladesh, Tanzania, Zambia, Poland, Togo, Swaziland, Kenya, Cameroon, Panama, Rwanda, Congo Dem. Rep., Chad, Guinea, Sierra Leone, Congo Rep., Bulgaria, Cambodia, Iceland, Vanuatu, Burundi, Lesotho, Honduras, Uganda, Malawi, Mauritania, Solomon Is, Iraq, Senegal, Angola, Haiti, Tonga, Trinidad and Tobago, Niger, Madagascar, Cuba, Zimbabwe, Nicaragua, Mongolia, Mozambique, Namibia, Botswana

Twenty-five years of success and failure in food production

● The graph gives an index number for average food production per head in the years 1994–96, taking the average of 1969–71 in all cases as 100. So it measures the overall change over a quarter of a century.

○ These data show how dangerous it is to generalize about the world. In something as basic as the production of the amount of food per head, countries could hardly differ more. At one extreme are countries which produce per person little more than half of what they produced twenty-five years ago, while at the other several countries have stepped up their food production per person by more than half during the same time.

It is notable from these figures that some of the countries which most increased their food production are the same as those which have experienced a particularly fast rate of economic growth in general. They are known as countries that have industrialized especially fast. Industrial growth has not seemingly been in conflict with agricultural growth. It may be indeed that there has been synergy between the two sectors, with one offering both inputs to and demand for the products of the other. Here perhaps is some confirmation of the virtues of sectorally balanced growth.

Special caution, however, is always needed in interpreting official figures for food production. They often leave out minor, non-market crops produced by peasants and smallholders. Due to the greed of some monocultures and the concentration of land holdings such production may fall when that of major cash crops rises.

■ World Bank 1999a.

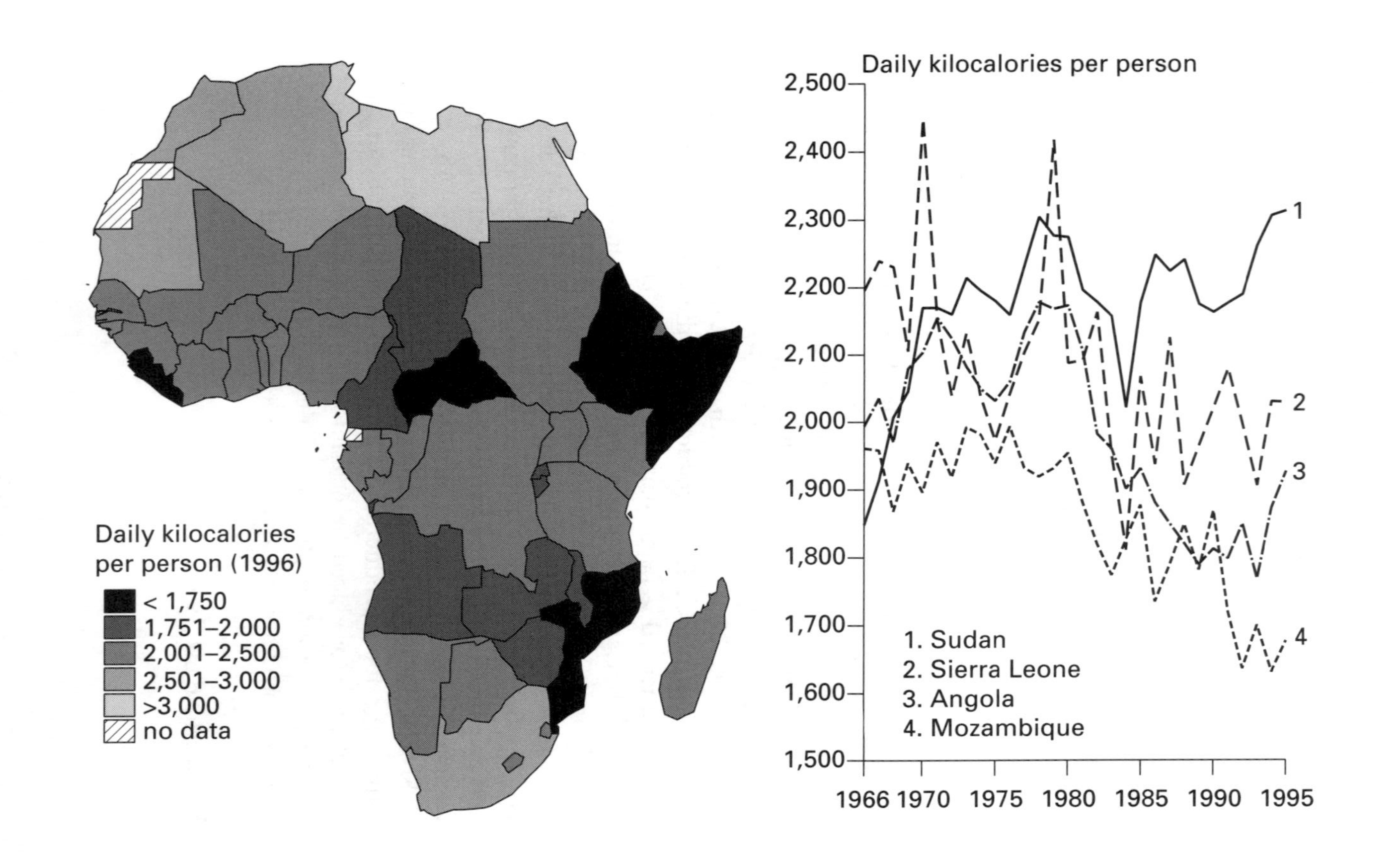
Daily kilocalories
per person (1996)
< 1,750
1,751–2,000
2,001–2,500
2,501–3,000
>3,000
no data
Daily kilocalories per person
2,500
2,400
2,300
2,200
2,100
2,000
1,900
1,800
1,700
1,600
1,500
1966
1970
1975
1980
1985
1990
1995
1. Sudan
2. Sierra Leone
3. Angola
4. Mozambique
1
2
3
4

● The map shows the implicit calorie supply in Africa in 1995 and the graph on the right shows the evolution of average calorie supply during the thirty years since 1966 for four African countries badly affected by civil war during this period.

○ The statistics of the FAO all point to Africa as the continent worst affected by problems of undernourishment and famine. Most of the countries with very low apparent calorie availability are African ones, although it is also true that the statistics are particularly defective and this has led some people to be sceptical of them. But there can be no doubt that widespread political strife in a number of African countries, and the resultant forced movements of population, have had disruptive effects on the agricultural situation. Africa is the continent with most displaced persons (› 100, 101). Food aid for forced migrants has often been insufficient to fill the gap and the continuation of the conflicts have in places (Sudan, for instance) made the distribution of food difficult. Extreme climatic effects, both droughts and floods, have also affected food production in a number of African countries. As the right-hand graph shows, three out of four countries suffering long civil wars have also experienced falling food energy supplies.

But it would be wrong to consider the whole continent indiscriminately as a nutritional disaster area. In spite of the problems mentioned, several African countries figure in the previous graph in the list of those which have lifted food production per head during the period from 1969 to 1997: Benin, Tunisia, Egypt, Morocco, Burkina Faso, Côte d'Ivoire, Central African Republic, Libya and Nigeria.

■ FAO 1997.

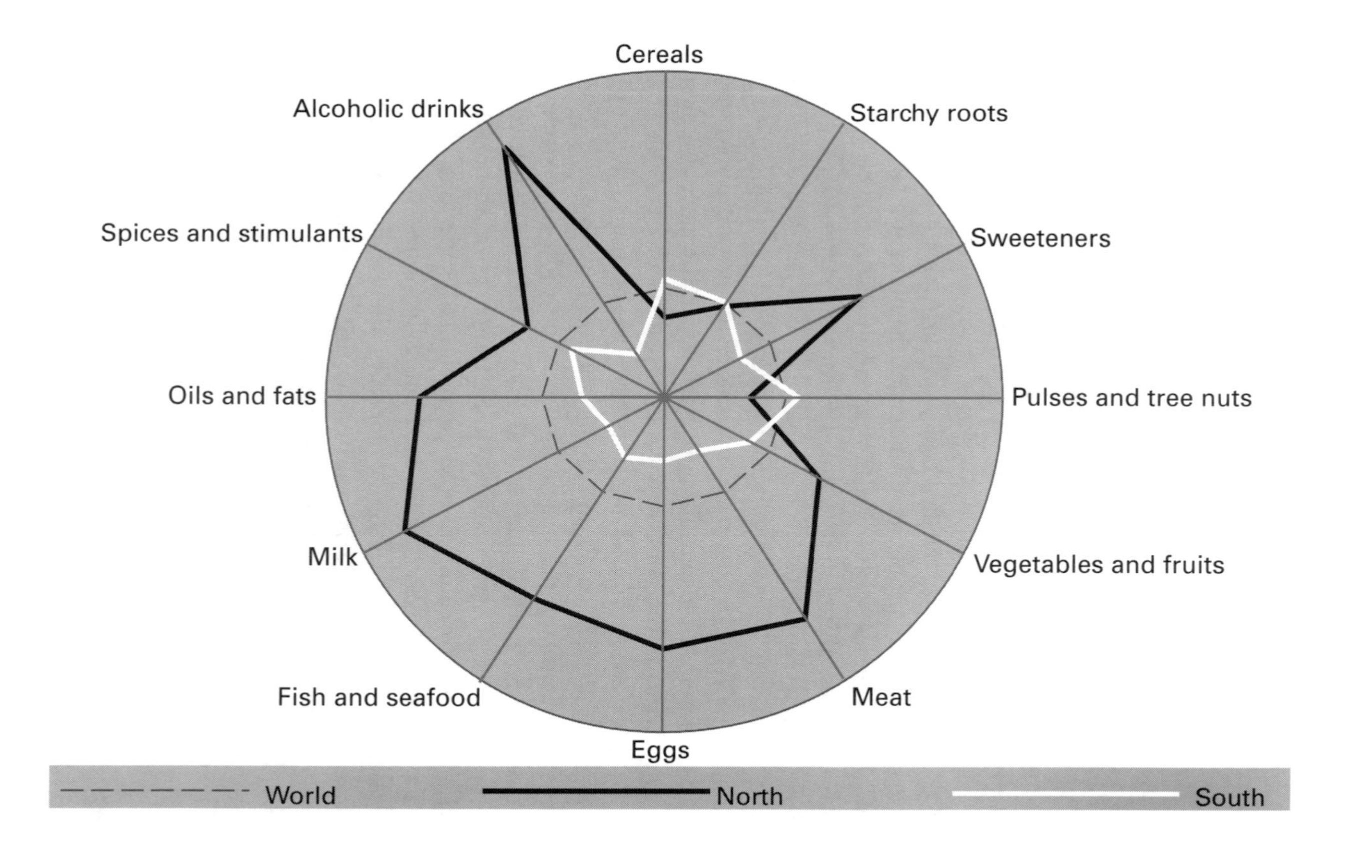
Cereals
Starchy roots
Sweeteners
Pulses and tree nuts
Vegetables and fruits
Meat
Eggs
Fish and seafood
Milk
Oils and fats
Spices and stimulants
Alcoholic drinks
World
North
South

● Another radial or radar chart describes the difference between the average diets of North and South. Each axis measures the consumption per head of a type of food in relation to the world average. The world consumption of each food (as indicated by number of calories) is set at a fixed distance along the axis so that world consumption appears as a regular dodecagon (twelve-sided figure). The dodecagons for North and South can then be compared with each other and with the world.

○ The division of the world into two parts is, as always, grossly oversimplified and conceals immense differences between the two groups. Yet as a step in understanding what the world eats it is extremely revealing. In the first place the different overall size of the diet pattern of the North and the South reveals that on average people in the South eat less of all kinds of food (in terms of calorie intake) than people in the North, with the exception of cereals, pulses and nuts. The average consumption of the other great staple food, starchy roots, is about equal. For all the other categories people in the North consume more, especially of proteins and fats of animal origin (› 55), of sweeteners and – the largest difference of all – of alcoholic beverages. The diet of the North, the developed diet, is sweeter, fatter, more plentiful, less vegetable, more meaty and more various than the diet of the South. While there is no doubt that the developed diet has many features which make it more attractive, questions abound about how much it really represents a gain in welfare and about the perverse effects of parts of it. Perhaps in relation to the international inequalities of diet, more clearly than in any other subject, it is relevant to speak not only of insufficiency but also of excess; not only of underdevelopment but also overdevelopment.

■ FAO 1997.

(for full labelling of the axes, see 51)

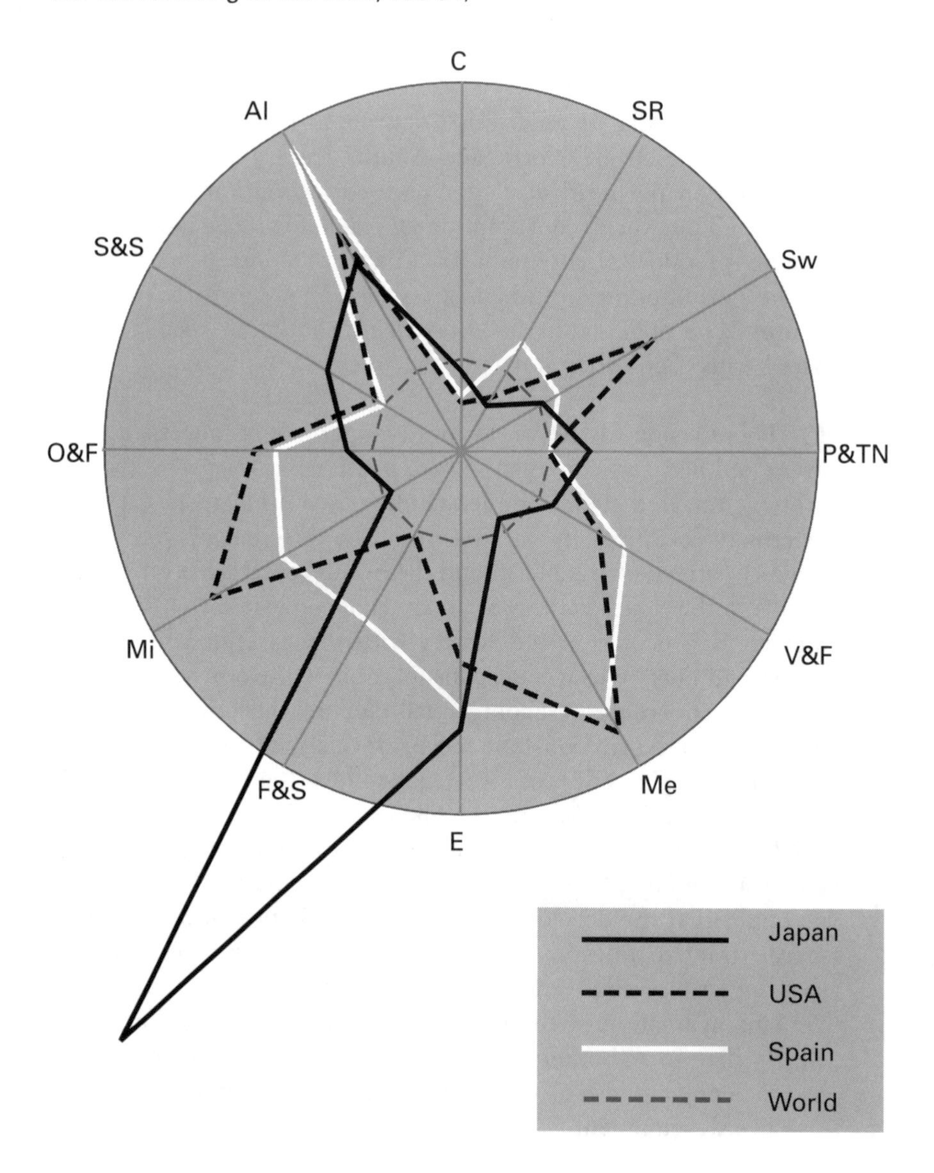

● This graph is the same as 51 with the labels of the axes abbreviated.

○ Although the comparison between North and South in the previous graph is revealing it suffers from the problem of all averages: that they conceal an unknown amount of divergence. This and the following graph, showing the average diets of three countries of the North and five of the South, reveal some of this divergence. Even these, of course, are only national averages and conceal yet more differences between people based on class, region and other variables.

The size of the dodecagon representing the world has been shrunk in this chart to take account of the fact that for these three well-fed countries the relative levels of consumption of some of the components are so high. What stands out here are the four peaks in the US diet: for sweeteners, meat, milk and alcoholic drinks. The high animal protein diet in the USA has partly been the result of the popular desire of an increasingly prosperous population for what are regarded as gastronomic luxuries. But not entirely. This diet has also been artificially encouraged during the twentieth century: by nutritionists who have advocated a diet high in animal protein as a route to maximum physical attainment and health; by those interested in military power, especially the US Army, on the grounds that large inputs of animal protein make strong soldiers (and also strong mothers of future strong soldiers); and by the very powerful lobby of cattle farmers and others in the meat industry. Some of this can be traced back to the reaction in Britain to the alarming rate of rejection of military recruits as physically inadequate during the Boer War.

■ FAO 1997.

(for full labelling of the axes, see 51)

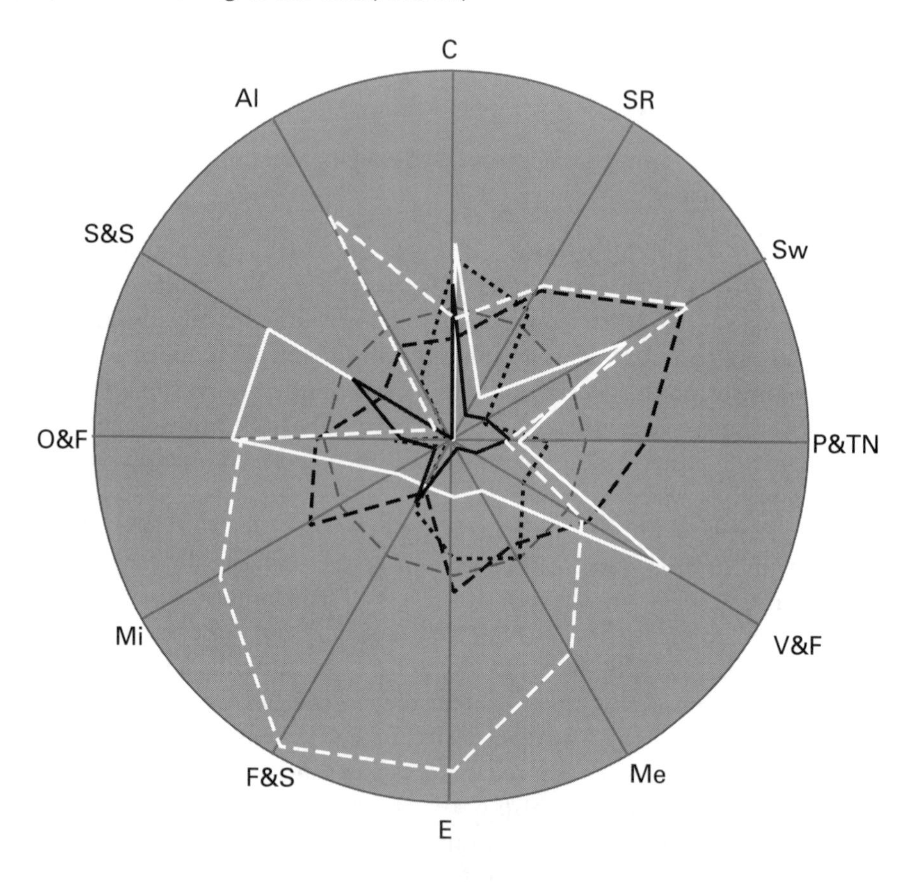

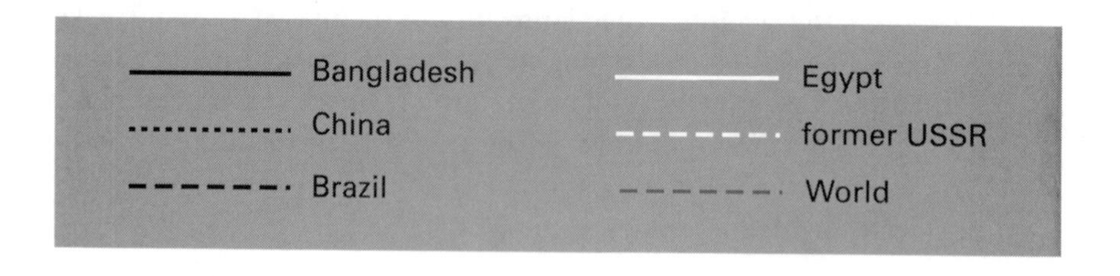

● This graph is also the same as 51 with the labels of the axes abbreviated.

○ Since so much of the diet of developed countries today has been influenced both by development itself and by example, it is not surprising to find that dietary patterns in the developing countries are even more variable. Some major variations are seen in this graph:

- the average Brazilian diet is similar to the world average but specially biased towards sweeteners and pulses;
- the average Russian eats a diet similar in structure to that of the USA but in smaller quantities, even of alcoholic beverages (possibly the result of the fact that in Russia drinking alcohol is largely confined to the male minority of the population);
- in all the other countries shown here alcohol consumption is much lower than the world average and in the case of Egypt (due to the influence of Islam) is virtually zero;
- the average Egyptian diet is an interesting one since it is one example of the Mediterranean diet (consisting of high intakes of cereals, fruit, vegetables, pulses and vegetable oil) now much extolled for health reasons by nutrition experts;
- the Chinese diet is remarkable for its relatively high content of cereal and starchy roots but it is also close to the world average in the consumption of meat and eggs;
- the diet of Bangladesh is most notable for its small size: consumption of all categories is well below the world average except for cereals (in this case rice) and spices. More will be revealed about this in the next chart.

■ FAO 1997.

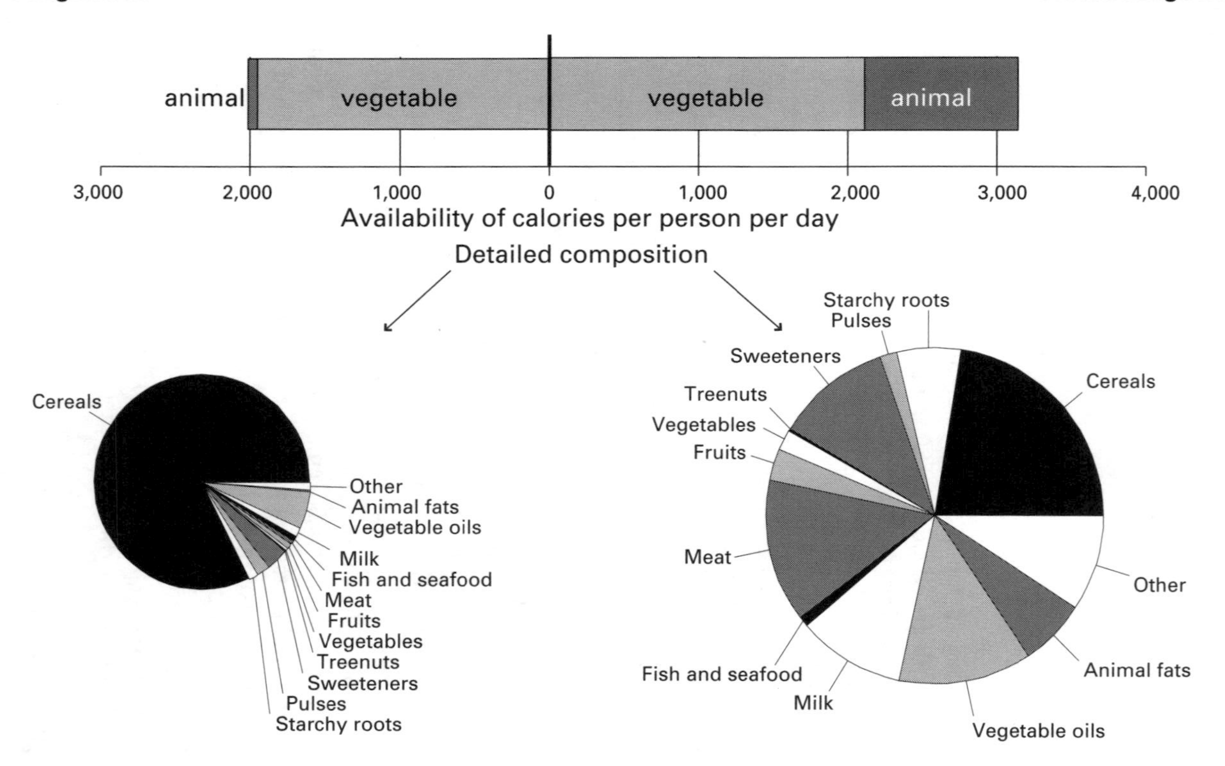
Bangladesh
United Kingdom
animal
vegetable
vegetable
animal
3,000
2,000
1,000
0
1,000
2,000
3,000
4,000
Availability of calories per person per day
Detailed composition
Cereals
Other
Animal fats
Vegetable oils
Milk
Fish and seafood
Meat
Fruits
Vegetables
Treenuts
Sweeteners
Pulses
Starchy roots
Starchy roots
Pulses
Sweeteners
Treenuts
Vegetables
Fruits
Meat
Fish and seafood
Milk
Vegetable oils
Animal fats
Other
Cereals

Quantity and variety: the diets of the UK and Bangladesh

● The top chart shows daily average calorie supply, divided into calories of animal and vegetable origin, for Bangladesh and the UK. And the pie charts below make a detailed comparison of the size and composition of the diets of these two countries.

○ Bangladesh is a poor country with widespread undernourishment and the UK a rich one with very little undernourishment. A number of points are revealed clearly by this comparison:

- in the UK the average calorie intake is more than 50 per cent higher than in Bangladesh;
- in the UK almost one-third of calories have animal origins; in Bangladesh the proportion is negligible;
- in Bangladesh over 80 per cent of the calories come from one category of food, cereals, more specifically rice, while in the UK cereals, once more important, now account for less than one-quarter of the calorie intake;
- in other words the British diet is extremely varied while the Bangladeshi diet is extremely monotonous.

The discussion of nutrition and undernourishment almost invariably focuses on quantitative adequacy (of energy, protein and micronutrients). It is a discussion within the field of medicine and development. Much less frequently food is treated also as a cultural phenomenon: its use in ritual, the pleasures of gastronomy, the joy of variety in available food. When it is said, as it frequently is, that enough food is available in the world to feed everyone, this is a statement only about the quantity of calories. Enough food may be available only if everyone eats a diet similar in structure, if greater in quantity, to that of Bangladesh. The inequalities in nutrition are about much more than numbers of calories.

■ FAO 1997.

The consumption of proteins...

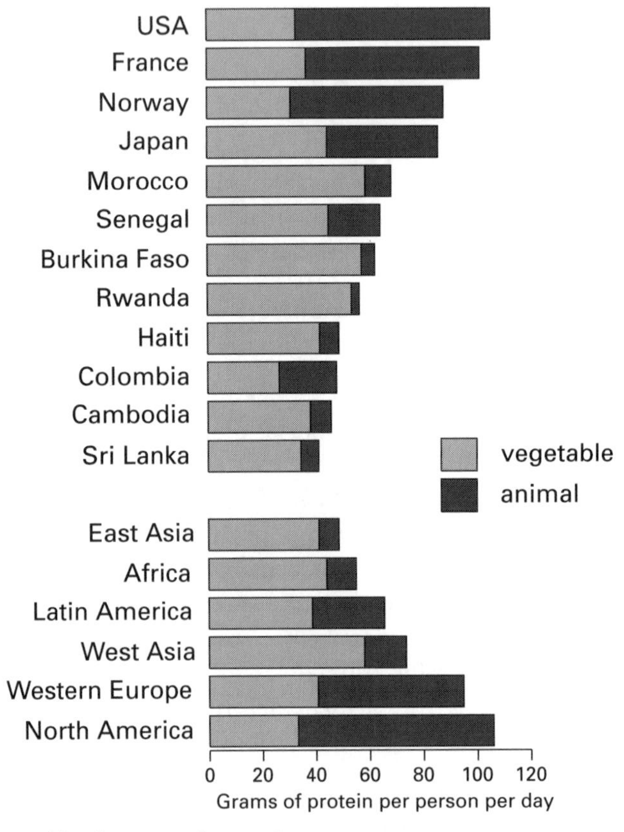

...reflected in the use of cereals

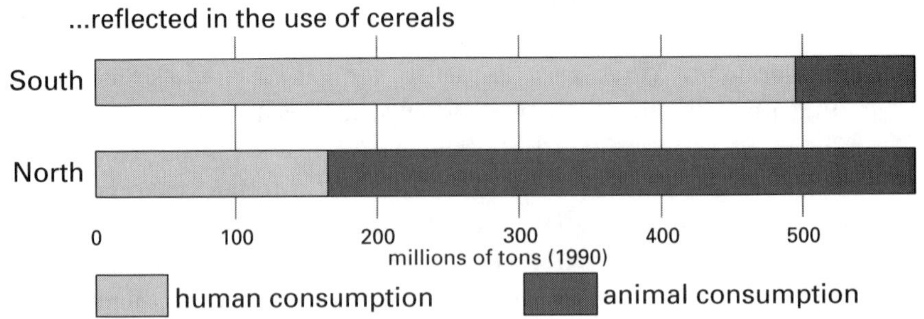

● The upper bar chart shows the average quantity and composition of daily proteins consumed in twelve differing countries and five world regions. The lower chart shows differences in the proportions of cereals used for animal and human consumption in developed and developing countries.

○ The quantitative differences in protein consumption between rich and poor countries are even more pronounced than the differences in calorie consumption. And, due to a combination of economic, cultural and religious factors, they are even more marked in the case of animal proteins. Nutritionists are less fervent than they once were about the need for high protein, and especially animal protein, consumption. Vegetarian diets are considered equally compatible with optimum health and in some respects are superior.

Meanwhile, however, the animals which humans in the North eat, themselves consume vegetable matter, at least when they are not forced to eat each other's brains. In the South cereals are eaten directly by humans, while in the North they are largely fed to animals. But every gram of animal protein requires about 10 grams of vegetable protein as input. By eating more meat humans have effectively put themselves on a higher level in the food chain at each stage of which about 90 per cent of energy is 'lost'. So the animal-protein-rich diet of the North requires much more land (and chemicals and machinery) per unit of food than the more vegetarian diet of the South. The diet of the North (like other aspects of the consumption levels of rich countries) could not be extended to the rest of the world because the resources simply do not exist. Perhaps it is time for a return of the spinach enthusiast Popeye, with his partner Olive Oyl.

■ FAO 1997.

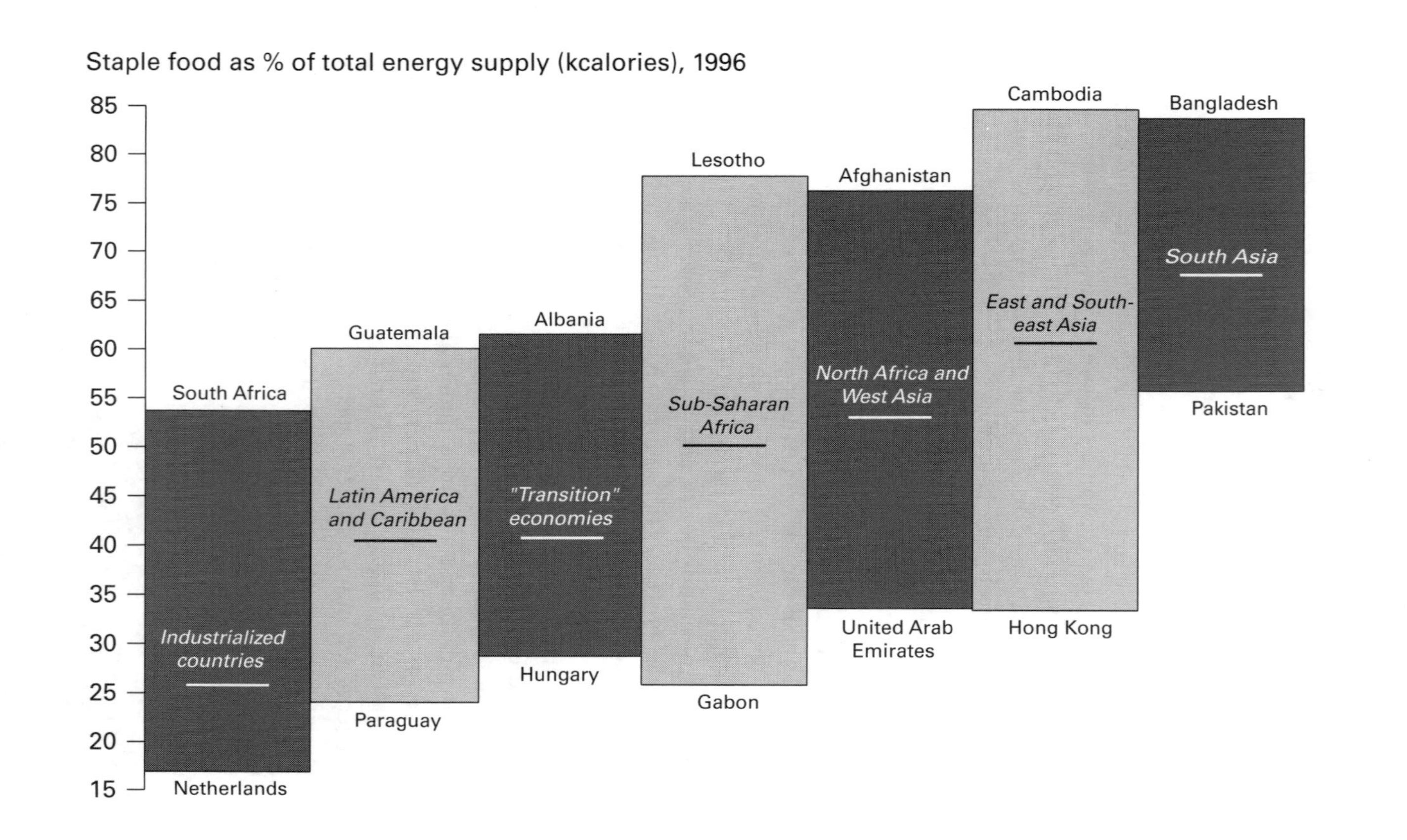
Staple food as % of total energy supply (kcalories), 1996
85
80
75
70
65
60
55
50
45
40
35
30
25
20
15
South Africa
Industrialized countries
Netherlands
Guatemala
Latin America and Caribbean
Paraguay
Albania
"Transition" economies
Hungary
Lesotho
Sub-Saharan Africa
Gabon
Afghanistan
North Africa and West Asia
United Arab Emirates
Cambodia
East and South-east Asia
Hong Kong
Bangladesh
South Asia
Pakistan

● The graph measures the proportion of a basic or staple food (usually cereals or roots) in the total food energy supply (kilocalories) of countries and regions. The line within each shows the proportion for the region as a whole, and the top and bottom of each bar are the values of the two extreme named countries in each region.

○ What we see here is the progressive elimination of the concept of the staple food with development. In South Asia 75 per cent of calories consumed come from cereals (rice or wheat depending on the part of the region). In industrialized countries this proportion has been reduced to less than one-quarter. With greater wealth staple foods come to be regarded as inferior foods to be replaced by greater nutritional variety and luxury. While it is true that staple foods can be monotonous, they have not been staples without reason. Their role is due to the fact that they can be produced without excessive resources and so can maintain life. They have been an economical way of converting human labour and natural endowments into the survival of the species.

Some nutritionists now doubt that greater variety in diets and unlimited choice of foods is good for health. They argue that many eating disorders and the qualitative inadequacy of diets may be encouraged by the inability to make rational nutritional choices in conditions of excessive variety. A staple food which produces regularity in the quantity and content of nutrition may, therefore, be within limits a nutritional virtue.

■ FAO 1997.

Four sources of inequality
sex, urban bias, regional differences and race

V

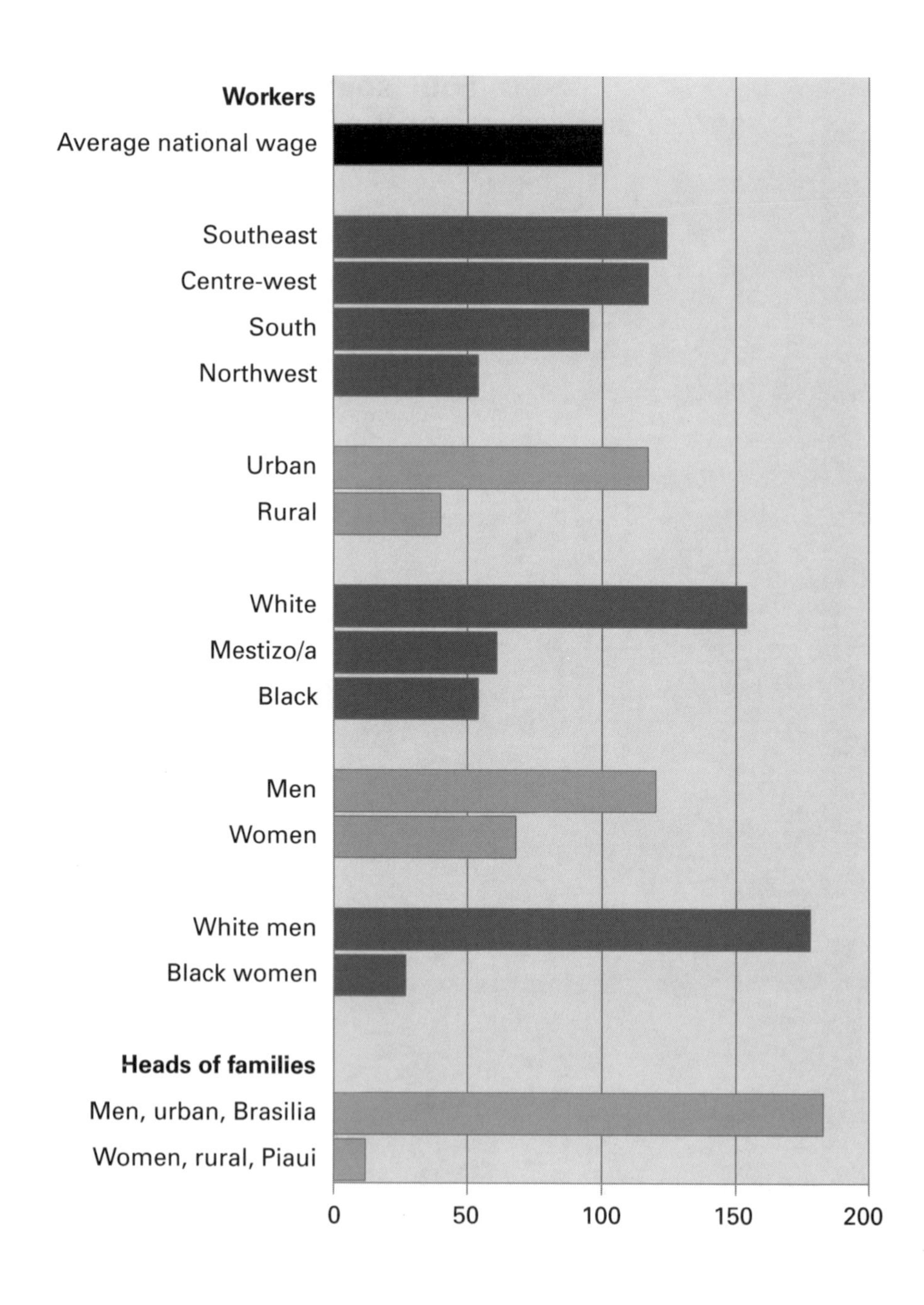
Workers
Average national wage
Southeast
Centre-west
South
Northwest
Urban
Rural
White
Mestizo/a
Black
Men
Women
White men
Black women
Heads of families
Men, urban, Brasilia
Women, rural, Piaui
0
50
100
150
200

Four sources of wage inequality: the case of Brazil

● This chart shows data on inequality in Brazil in the early 1990s. If the average national wage is set at 100, the other five sets of bars show the difference in average pay from the national figure according to region of the country, urban or rural residence, colour, sex, sex and colour combined, and finally sex and urban/rural residence combined.

○ Brazil's exceptional inequality (‹10) is caused, as in many countries, by a multiplicity of discriminations. But unlike many countries, Brazil publishes a large amount of information about different kinds of inequalities among its people. This chart summarizes some of that information:

- wages in the southeast of the country are about 2½ times those in the northwest (›64);
- urban wages are about 3 times rural wages;
- wages of whites are about 3 times wages of blacks;
- and wages of men are nearly double those of women.

Real inequalities are wider because they result from a combination of overlapping discriminations. So a white man typically earns 7 times the pay of a black woman and an urban male head of family in a rich region earns 15 times as much as a female rural head of family in a poor region. These data provide a very good example of the multiple sources of inequality – class, sex, place of residence and colour – which are examined one by one in the rest of this section of the book. The data, however, are only for wage and salary earners in employment. Capitalists, landowners, property owners on the one hand, and unemployed people, peasants, self-employed and indigenous people on the other are all missing. Their inclusion would explain more of Brazil's exceptional inequality.

■ Fundacão Instituto Brasileiro de Geografia e Estatística 1994.

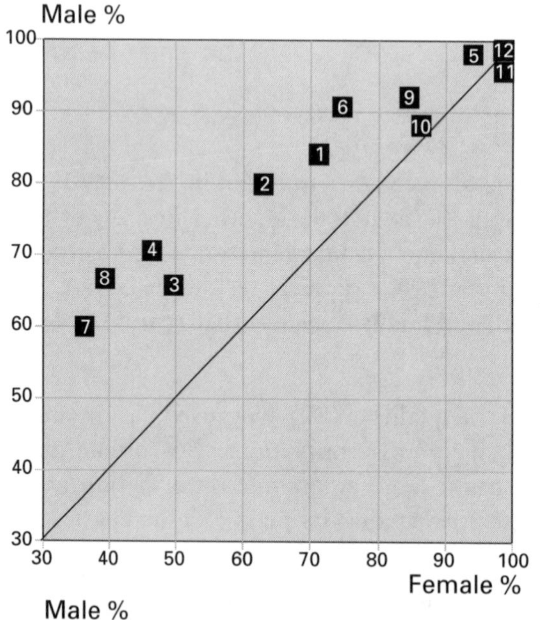

Relation of male and female adult literacy rates (1997)

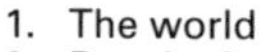

1. The world
2. Developing countries
3. Sub-Saharan Africa
4. Arab States
5. East Asia (except China)
6. China
7. South Asia (except India)
8. India
9. Southeast Asia and Pacific
10. Latin America and Caribbean
11. Eastern Europe and CIS
12. Industrialized countries

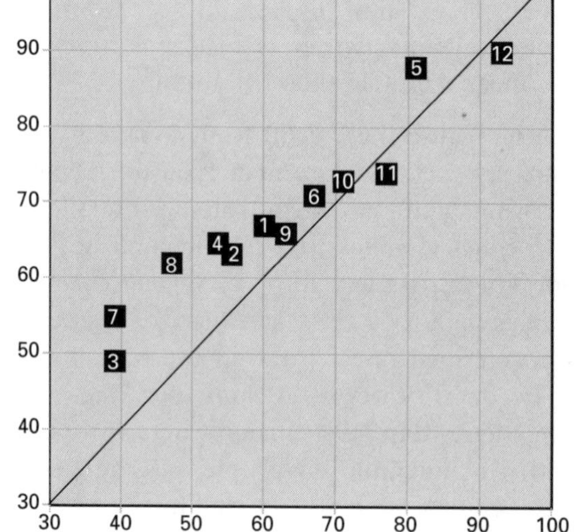

Relation of male and female combined educational enrolment ratio (1997)

Sexual bias in access to education, 1997

● The two graphs show aspects of sex discrimination in education. The vertical axes measure literacy and educational enrolment levels of men, the horizontal axes the same variable for women. The numbered points show the values for the world and eleven subcategories of country. If men and women were equal these points would be on the 45° line; points above show discrimination against women. Combined educational enrolment is an index compiled by the United Nations Development Programme, based on the percentage of the appropriate age group enrolled in primary, secondary and tertiary education.

○ The graph shows differences by region and category of country in both the level of literacy and enrolment and the differences between men and women. The developed countries, Eastern Europe and the former USSR and East Asia have very high, in many cases effectively complete, levels of literacy. So it is logically impossible for there to be differences between men and women. But for all other groups of countries, except Latin America, women remain considerably behind men in their access to reading and writing skills. The differences are more severe than in the case of present school enrolment. This is because adult literacy differences express past differences in access to education which were even more discriminatory than those of today. Nevertheless, with the same exceptions as before, women in most areas of the world still find it more difficult than men to gain access to education. The difference is most acute in the Arab countries, India and the rest of South Asia, part of a story which we have seen and will see other parts of elsewhere in this book (‹ 26–33, › 123).

■ UNDP 1999.

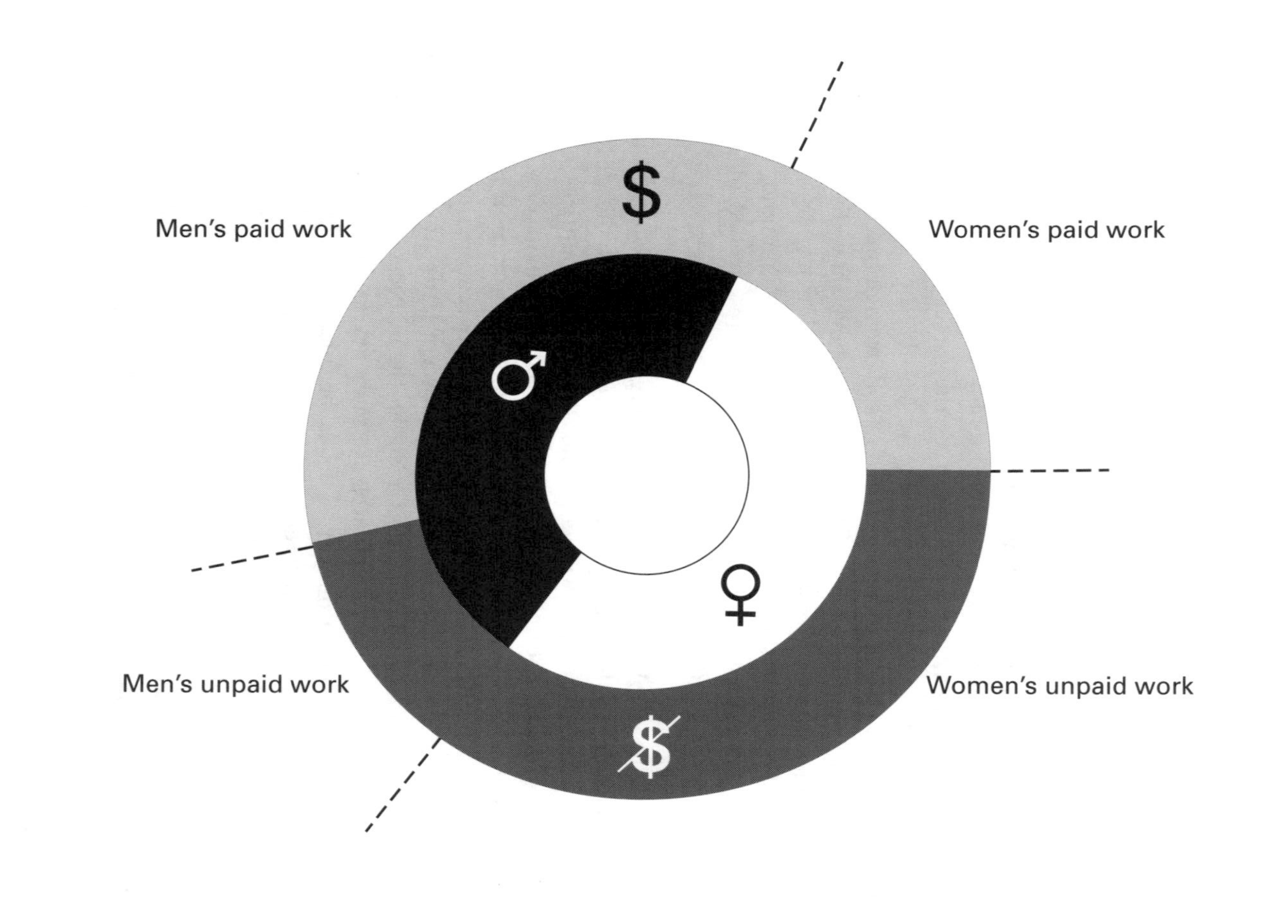
Men's paid work
$
Women's paid work
♂
♀
Men's unpaid work
Women's unpaid work
$

The sexual bias of work and pay, 1995

● The graph consists of two rings which divide the total hours worked in the world according to two criteria: whether the work is done by men or women and whether it is paid or unpaid. The rings are then placed in relation to each other to show the four categories which result from the two criteria combined: men's and women's paid and unpaid work.

○ In 1995 the Human Development Report published the results of an original study about the world sexual division of labour. Among other things, it concluded:

- women did 53 per cent of the total hours of work in the world and men 47 per cent;
- about 75 per cent of men's work is paid;
- only about one-third of women's work is paid;
- so, of the four categories of work, the largest is women's unpaid labour and the smallest is men's unpaid labour.

These results are based on generalizing a number of sample studies which have been carried out in individual countries. They are, therefore, quite tentative but since they all tell a similar story it is justifiable to generalize them.

This difference in the nature of women's and men's work is the result of the discriminatory nature of gender roles and also partly explains why changing these roles is difficult. Women work much less than men for pay; and when they do they work for less pay. The combination of these two facts leads to men taking a very large direct share of monetary income in nearly all societies and impedes women's economic independence (›61).

■ UNDP 1995.

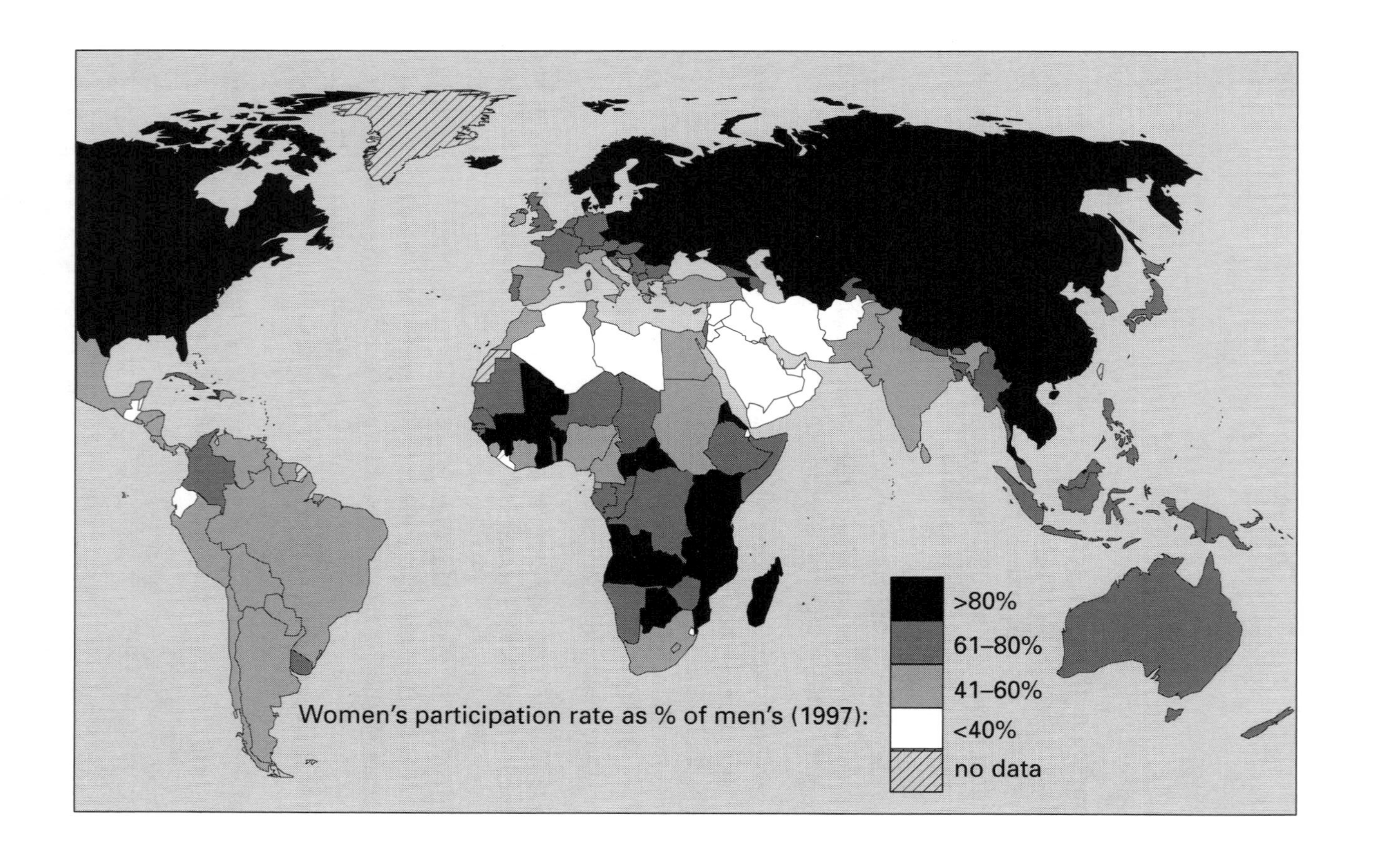
Women's participation rate as % of men's (1997):
>80%
61–80%
41–60%
<40%
no data

Women and economic activity, 1997

● The map shows women's economic activity rate as a percentage of men's. The rate of participation in economic activity is officially defined as supplying labour for the production of goods and services. In principle it includes self-employed people and unemployed people who wish to work. It may also include those who work outside the home in unpaid community work or in unpaid agricultural labour, but different countries apply differing criteria. It is not supposed to include unpaid domestic labour.

○ There is much variation in this indicator. Part of it may be due to the fact that national concepts of economic activity may differ and so the measure is not a very satisfactory one. But much of the variation is due to vast differences in social customs relating to the relative roles of men and women in society, and to differences in policies which facilitate or impede women's economic participation. It is noticeable that the highest rates of participation are in north-west Europe, North America and the formerly or presently communist countries. In the developed capitalist countries women's participation in paid labour has expanded enormously during the years since 1950. In some countries it has been facilitated by changing attitudes and by measures such as the provision of child-care and new laws on maternity leave. But another major cause has been the financial pressure on working- and middle-class families in countries, including the USA, with little growth of male wages (‹ 16).

Lower women's participation rates are found in the countries of southern Europe, parts of Asia and all of Latin America. But the lowest, not surprisingly, are in countries where conservative forms of Islam and Christianity are powerful.

■ UNDP 1999.

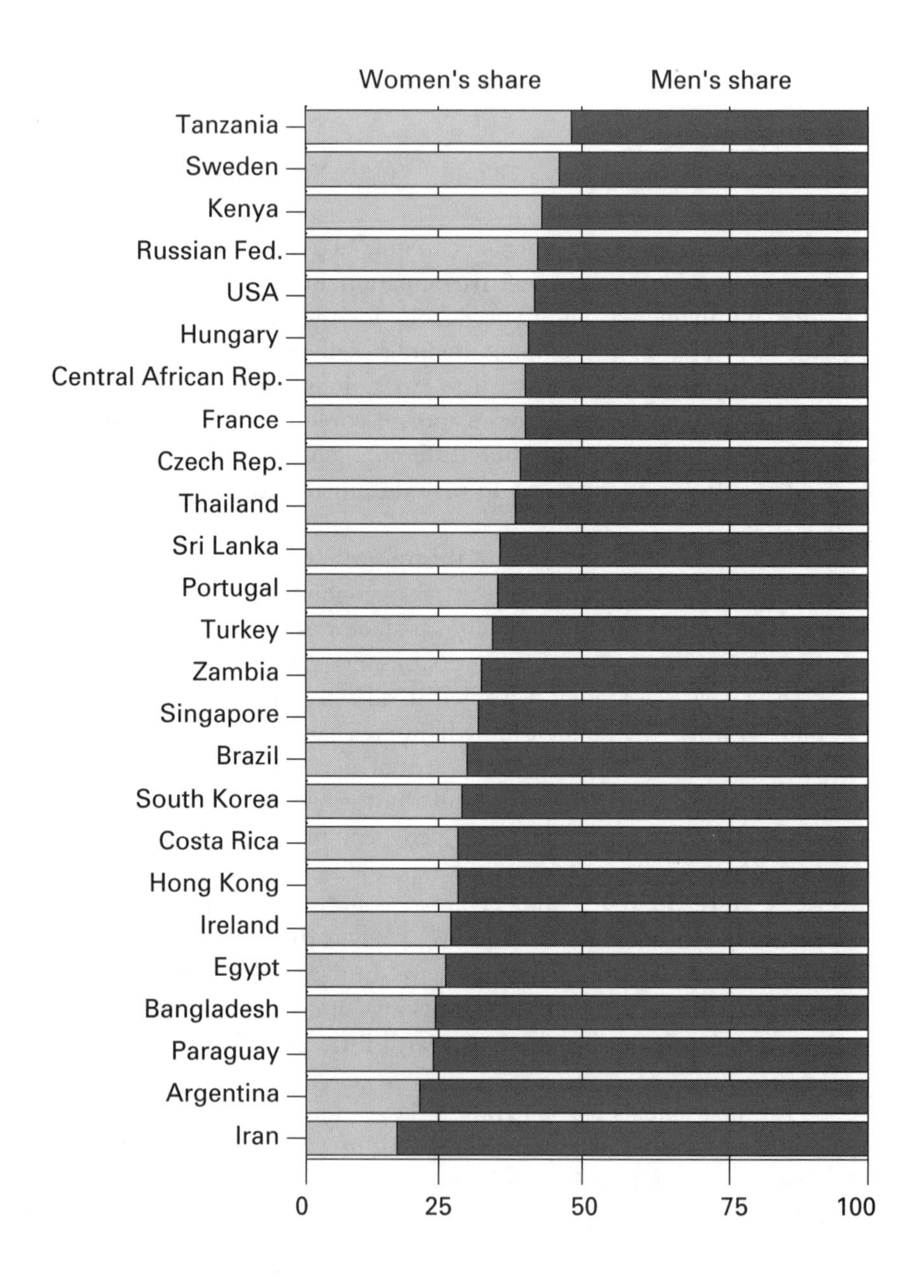
Women's share
Men's share
Tanzania
Sweden
Kenya
Russian Fed.
USA
Hungary
Central African Rep.
France
Czech Rep.
Thailand
Sri Lanka
Portugal
Turkey
Zambia
Singapore
Brazil
South Korea
Costa Rica
Hong Kong
Ireland
Egypt
Bangladesh
Paraguay
Argentina
Iran
0
25
50
75
100

Women's and men's share of earnings, 1993

● The graph shows for a varied sample of countries the division of total personal monetary income between men and women. This depends on relative women's participation and relative women's pay.

○ The country where women's share of income is the highest is Tanzania. As in many African countries women's participation, both in traditional agriculture and in the commercial monetary economy, has traditionally been extremely high, producing relatively high independent incomes. In the case of Tanzania it has been further boosted by the ideology of its government since independence. In second place in this sample is Sweden. In all Scandinavian countries women tend to participate in the economy and in politics at a greater level of equality with men than in other Western countries. After another East African country (Kenya) comes Russia, where the Soviet legacy of high women's participation, officially encouraged by state ideology and policies, persists. Pay has been relatively equal for equal work, and women occupy many professional posts, though not so many responsible managerial ones. Next comes the USA, where relatively high women's income is relatively recent, and has resulted from an enormous increase in the number of women participating in the labour market.

At the other end of this sample are a number of countries where the combination of social tradition and religious teaching contribute to obstacles to women having an independent income. These countries tend to be in West Asia and Latin America. No sub-Saharan country is in this category and, with the exception of Ireland, no European country either.

■ UNDP 1996.

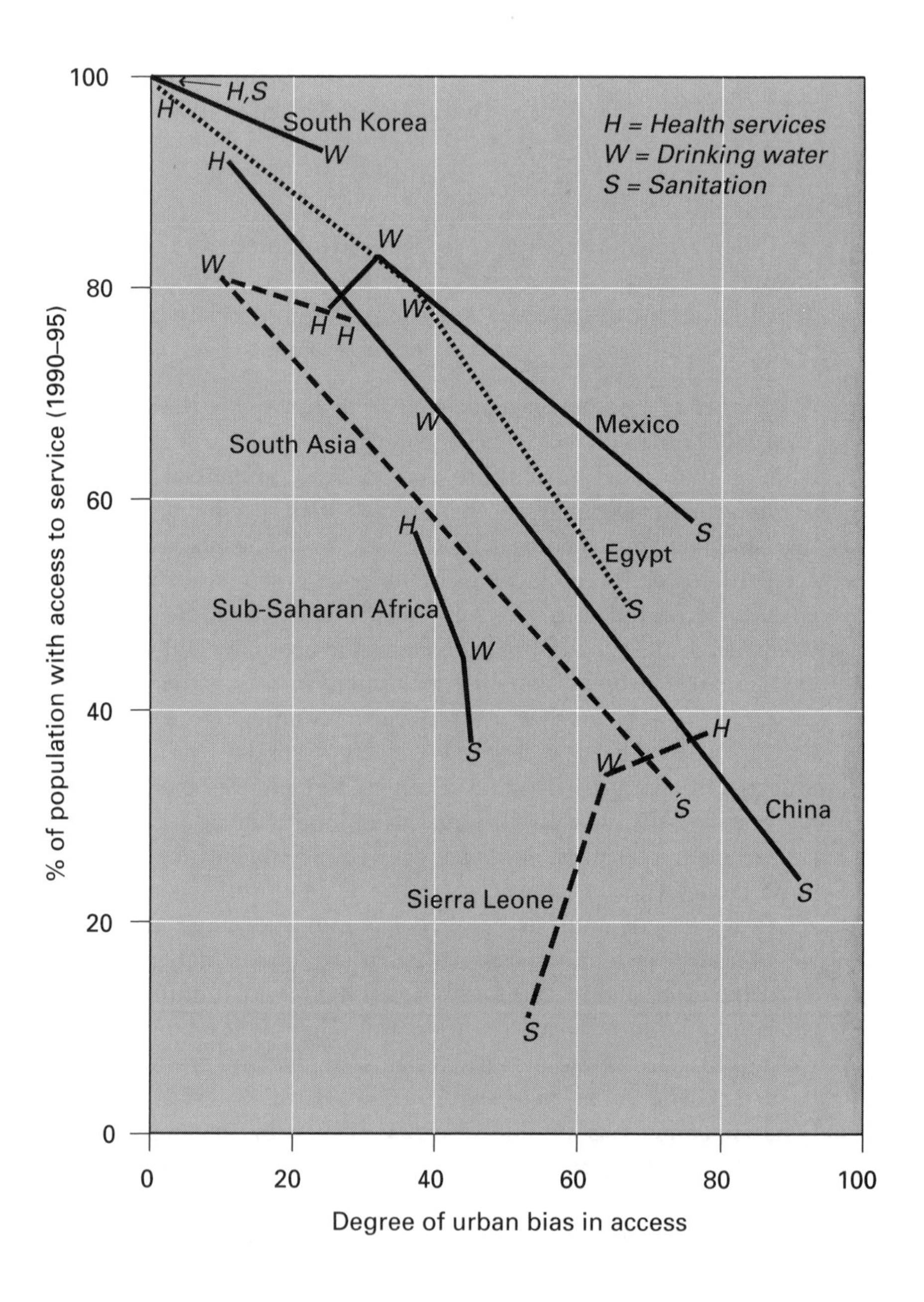

H,S
H
South Korea
W
H = Health services
W = Drinking water
S = Sanitation
H
W
W
W
H
H
W
Mexico
South Asia
60
S
H
Egypt
Sub-Saharan Africa
S
W
S
H
W
S
China
S
Sierra Leone
S
100
80
60
40
20
0
% of population with access to service (1990–95)
0
20
40
60
80
100
Degree of urban bias in access

● This graph describes the degree of urban bias in five countries and two regions. On the vertical axis is measured the percentage of the population as a whole having access to health services (H), drinking water (W) and sanitation (S). And the horizontal axis shows the degree of urban bias measured as the excess of the urban rate of access over the rural rate. If the percentage of the rural population with access is the same as the urban then this figure is zero. If urban access is 80 per cent and rural access is 20 per cent then the urban bias index is 60. For each country/region the three points, H, W and S, are linked by a line which provides a kind of footprint describing relative access to public health benefits.

○ What is noticeable from this graph is that in most countries access to health services is the highest of these three indicators and sanitation is the lowest. The differences in some cases are extreme; in China, for instance, over 90 per cent of people have access to medical services while less than 25 per cent have sanitation. Since most of the lines slope downwards from left to right this implies that the lowest indicator (usually sanitation) shows more urban bias than the others. Again in the case of China rural access to health services is only 10 per cent less than urban access. But in the case of sanitation the difference is 95 per cent. If this figure is accurate it really means that urban sanitation is general while rural sanitation is virtually unknown. Only those who have had to live without sanitation will know what a hardship this is, producing inconvenience, indignity and danger to health. The author of this book chooses to live in not very sophisticated rural conditions; but he would not be so happy to do so in the absence of sanitation!

■ UNDP 1999.

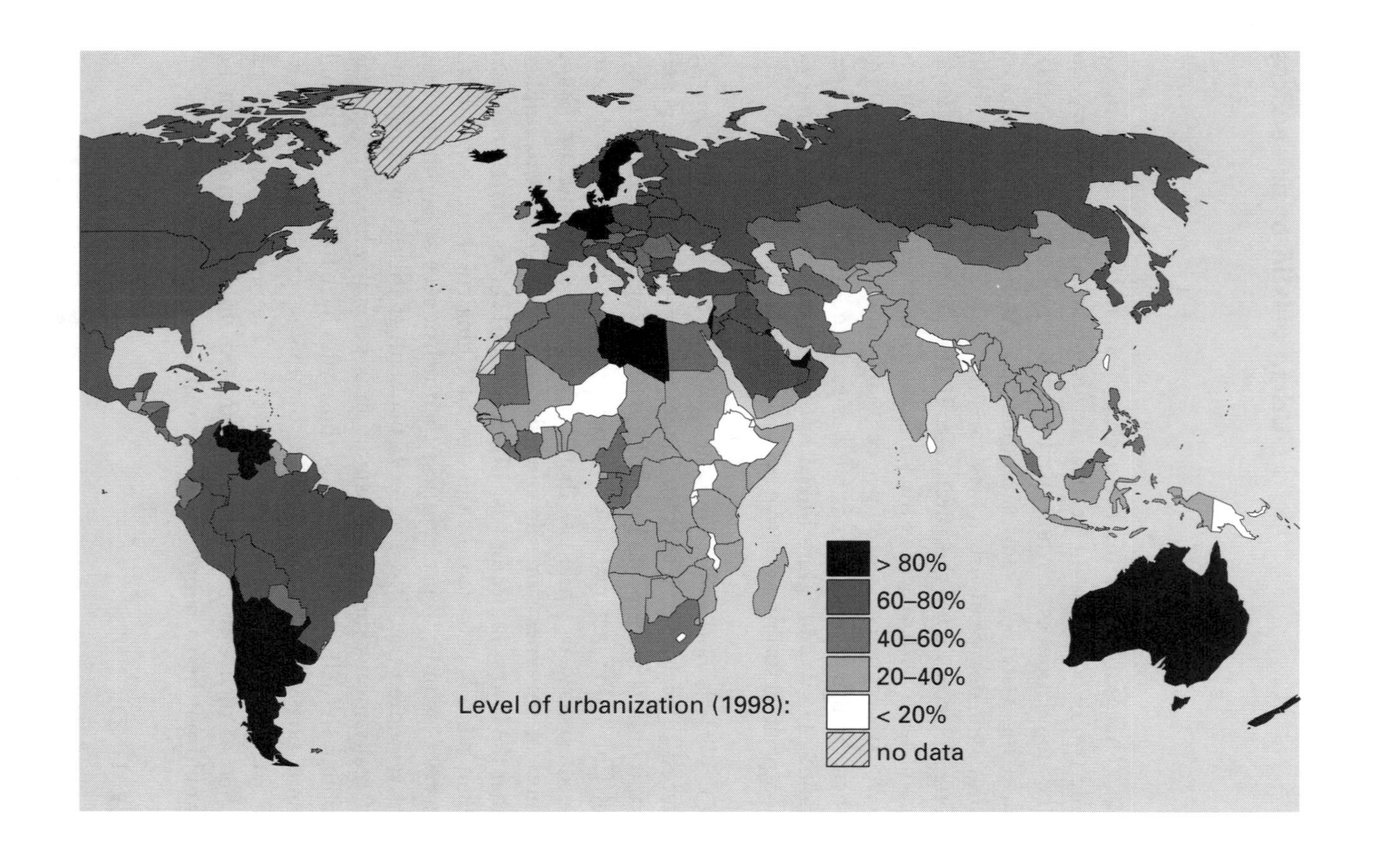
Level of urbanization (1998):
> 80%
60–80%
40–60%
20–40%
< 20%
no data

● The map shows the percentage of the population living in urban areas. The comparison should be treated with much caution since different countries use different definitions of the minimum population in an urban settlement (varying between about 1,000 and 10,000).

○ Urbanization has been part of the process of industrialization during the last 200 years. It is associated with the great productive transformation from a world in which nearly everyone produced their own food (along with a surplus for the minority of urban dwellers and for the rulers and parasites) to one in which the majority work in industry or services and are supplied with food from a minority who work in agriculture and live in rural areas (‹ 2–5). But this transformation has taken place in some parts of the world much more fully than in others. And urbanization in some places is a response not so much to the opportunities provided by new economic activities as to the impossibility of continuing to survive in the old ones. Nonetheless it is now estimated that around half of the world's population live in towns. The almost universal discrimination against the population of rural areas, which we saw in the previous graph, is therefore something from which 50 per cent of the world's population may suffer.

The problem of resolving urban bias derives from the fact that political and financial power tend to be concentrated in the urban parts of the world. This is not to say that large parts of the urban population are not oppressed; but the oppressed rural population tends to suffer a double burden of discrimination. And they have less opportunity to mobilize political power than their urban counterparts.

World Bank 1999a.

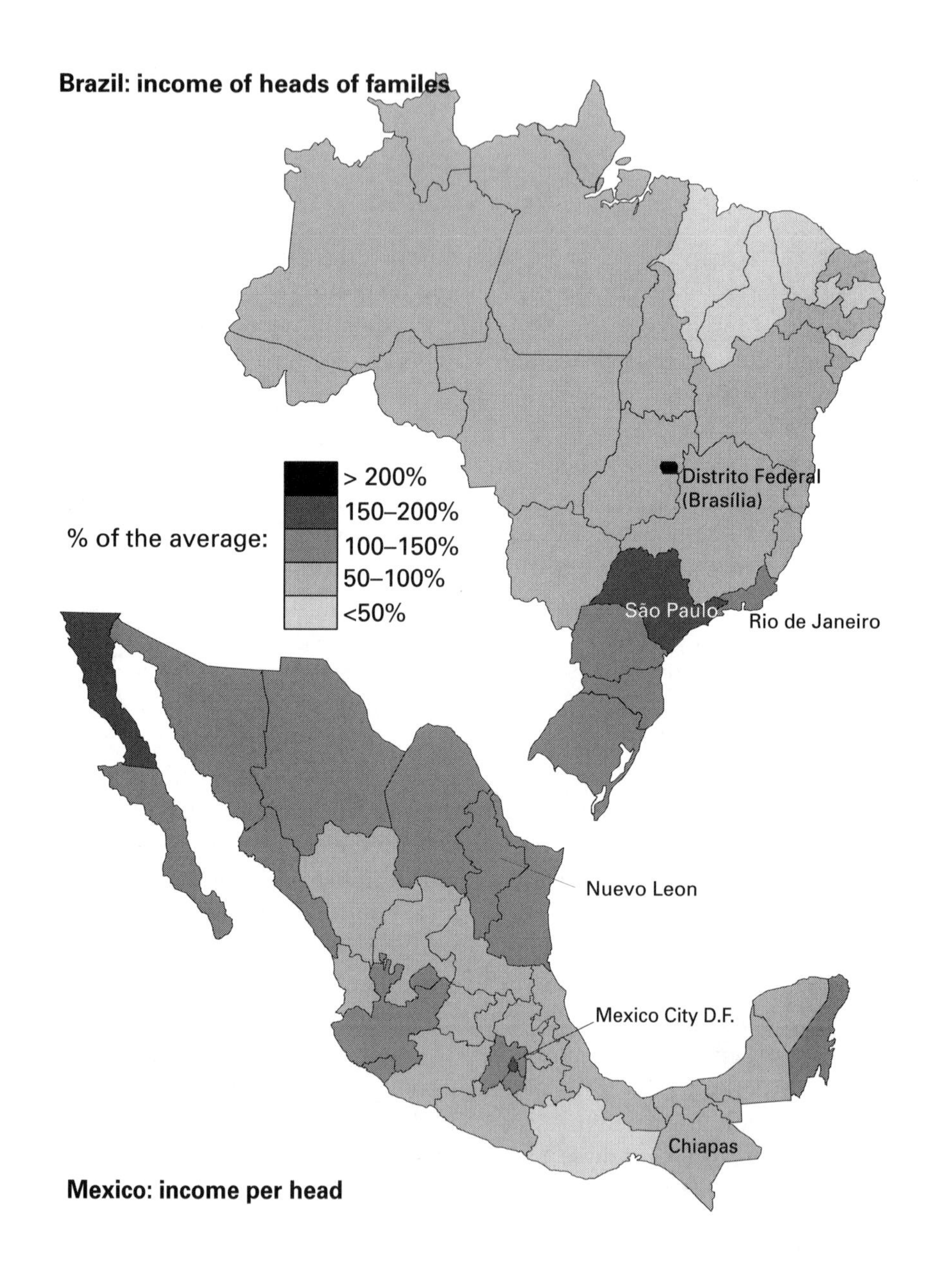
Brazil: income of heads of familes
% of the average:
> 200%
150–200%
100–150%
50–100%
<50%
Distrito Federal
(Brasília)
São Paulo
Rio de Janeiro
Nuevo Leon
Mexico City D.F.
Chiapas
Mexico: income per head

Regional differences in Brazil and Mexico

● This and the next three pages of maps show variations in the level of income between regions, states or provinces of seven countries and the European Union.

○ If the countries of the world are unequal then so are the regions of any given country; and regional inequality is apt to be more extreme in countries of the South than in countries of the North. One juridical country, especially if it is large, can encapsulate several social and economic countries.

Inequalities between the provinces of Brazil result from a development process which is extremely concentrated in the south-east of that vast country. The figures for income per head and for other indicators in the province of São Paulo are very close to those of a developed country, while in the northeastern provinces they often are no higher than those of a poor African country. The UNDP has disaggregated its Human Development Index for Brazil. The national average of the Index is 0.756 but regional values range between 0.831 for the south of the country (similar to Portugal or Singapore) and 0.544 in the northeast (similar to Bolivia).

In Mexico the inequality between states is also considerable, although less than that of Brazil. From the map it is evident that a major explanatory variable for differential levels of income per head is proximity to the USA. The inequality is perhaps less 'autonomous' than Brazil's. The disaggregation of the Human Development Index for Mexico (where the national average is 0.804) gives values ranging between 0.868 (similar to the Baltic republics) in the state of Nuevo León and 0.627 (the level of Jordan) for the state of Chiapas.

■ UNDP 1996; Demosphere International/Microsoft 1999; Fundacão Instituto Brasileiro do Geografia e Estatística 1994.

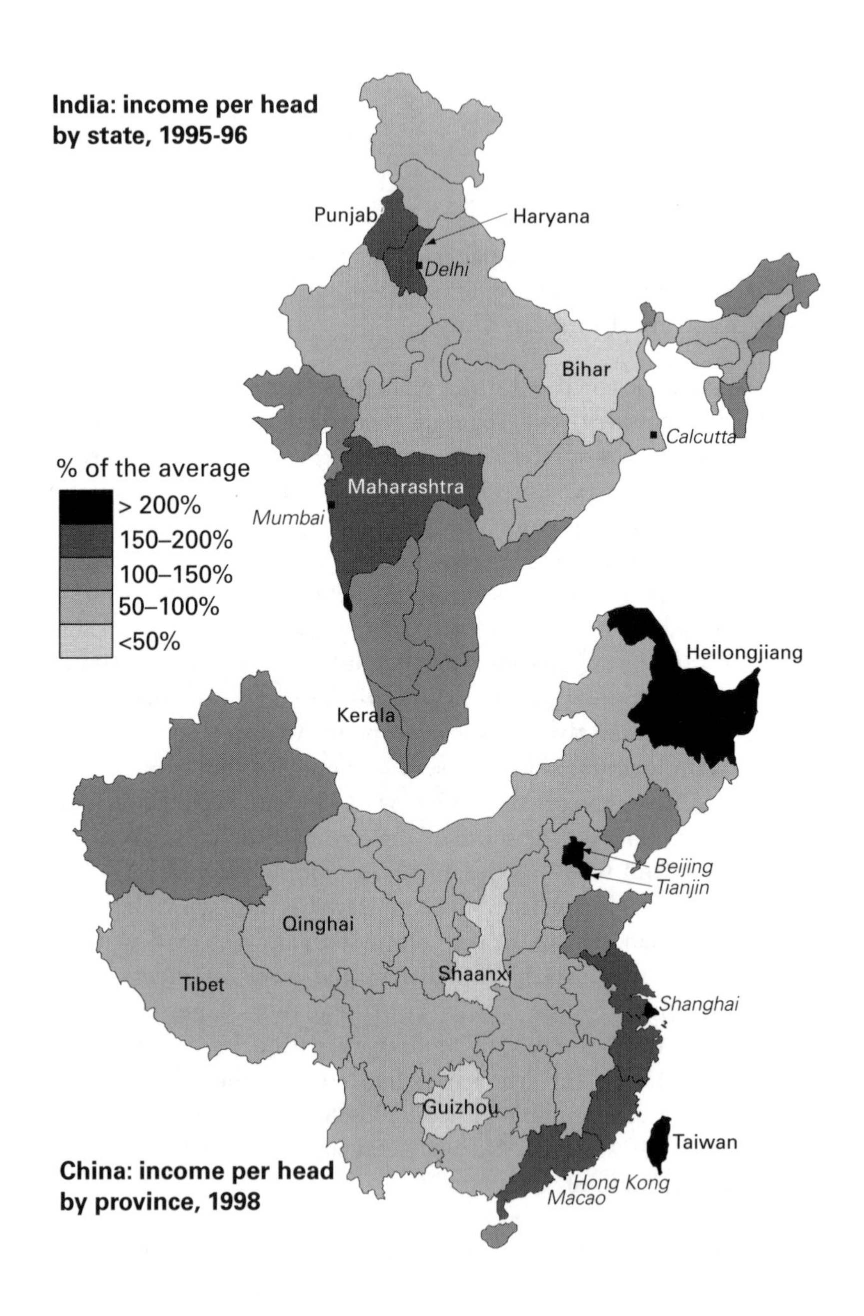
India: income per head by state, 1995-96
Punjab
Haryana
Delhi
Bihar
Calcutta
% of the average
> 200%
150–200%
100–150%
50–100%
<50%
Maharashtra
Mumbai
Kerala
Heilongjiang
Beijing
Tianjin
Qinghai
Tibet
Shaanxi
Shanghai
Guizhou
Taiwan
Hong Kong
Macao
China: income per head by province, 1998

Regional differences in India and China

● The maps of India and China show the levels of income per head for states and provinces as a percentage of the national average.

○ India and China are the most populated countries of the world, both with more than 1,000 million inhabitants, and both cover a massive geographical area. It is not, therefore, surprising that they both encompass huge regional differences. In China the inter-provincial range of income per head between Shanghai, the richest province, and Guizhou, the poorest, is almost 12 to 1. These income levels are close to those of Spain and Sudan respectively. In general the coastal regions have the highest incomes.

In 1995 the UNDP estimated the national value of the Human Development Index in China as 0.644, ranging from Shanghai with 0.885 to Guizhou with 0.494 and Tibet with 0.391, equivalent to the range between Argentina and Togo. China is in this sense many countries and these differences have grown in the last twenty years of extremely rapid economic growth.

The variation in income per head between Indian states is slightly less than that between Chinese provinces. In 1995–96 Delhi and Goa were the richest regions; the richest large state was Punjab, followed by Maharashtra, and the poorest was Bihar. The ratio between the average income of Delhi and Bihar was 6.2 to 1, and their levels were equivalent to those of Guatemala and Rwanda. The interstate variation in the Human Development Index does not coincide with the variation in income. Kerala, a state with only half the per-capita income of Punjab, had the highest HDI (0.6279) compared to Punjab's HDI of 0.5486. Bihar's HDI, however, was lower than that of any country in the world.

■ Surfchina 2000; UNDP China 1999; India Profile 2000; Economy Watch 2000; UNFPA 1997.

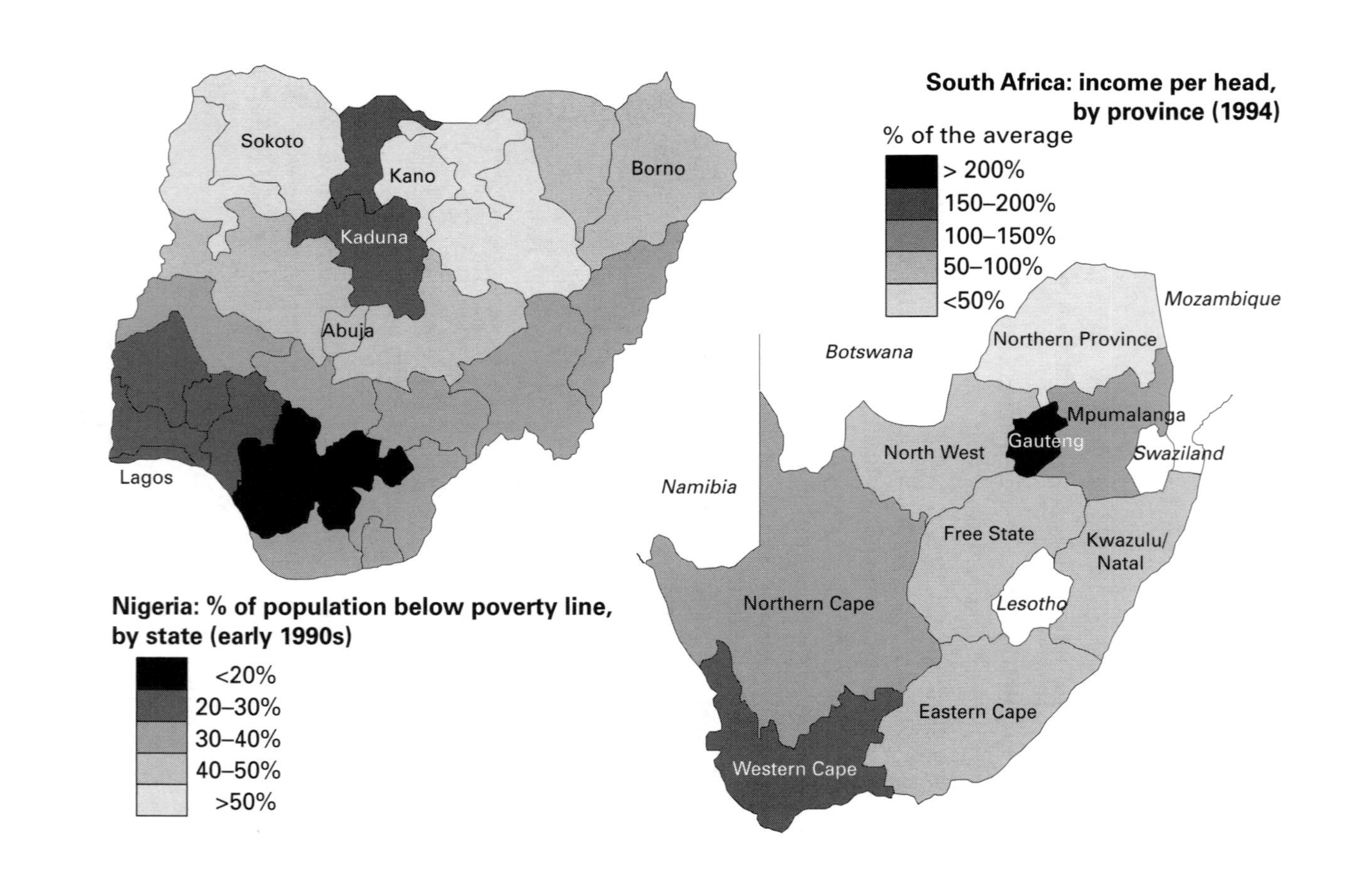
Sokoto
Kano
Borno
Kaduna
Abuja
Lagos
Nigeria: % of population below poverty line, by state (early 1990s)
<20%
20–30%
30–40%
40–50%
>50%
South Africa: income per head, by province (1994)
% of the average
> 200%
150–200%
100–150%
50–100%
<50%
Mozambique
Northern Province
Botswana
Mpumalanga
Gauteng
North West
Swaziland
Namibia
Free State
Kwazulu/
Natal
Lesotho
Northern Cape
Eastern Cape
Western Cape

Regional differences in Nigeria and South Africa

● For Nigeria the map is based on a study of the interstate prevalence of poverty, since state income per head figures are not available. For South Africa the data are for relative income per head by province as in the previous maps.

○ South Africa has, as the legacy of apartheid, a level of inequality almost unsurpassed in the world (‹ 10). Vast differences in economic level exist between ethnic groups, social classes and, as is plain here, between regions. In 1994 income per head ranged between 14,737 rands in Gauteng (which includes Johannesburg) and 1,944 rands in Northern Province, a ratio of 7.6 to 1, equivalent to the range between Argentina and Nepal.

Nigeria is another African country with wide regional differences. They are shown here as variations in the level of absolute poverty. These show a phenomenon which is general in West Africa: that the coastal zones are more developed than those of the interior. The 1992/93 poverty level ranged between 14.4 per cent in Imo and 55.8 per cent in Bauchi, with a national average of 34.1 per cent. In 1994 Nigeria's Human Development Index was 0.348, a figure which puts it close to the bottom of the scale. A study by the UNDP of interstate differences showed a range between 0.666 for Bendel (higher than the national level of Cuba in that year) and 0.156 in Borno (lower than the national level of any country). Life expectancy was 39.6 years in Borno and 59.5 in Bendel. The adult literacy percentage for the same states was 12.1 per cent and 79.5 per cent, respectively. These are wider differences than for any other country mentioned in this group of maps.

■ World Bank 1995b; *Financial Times* 1995.

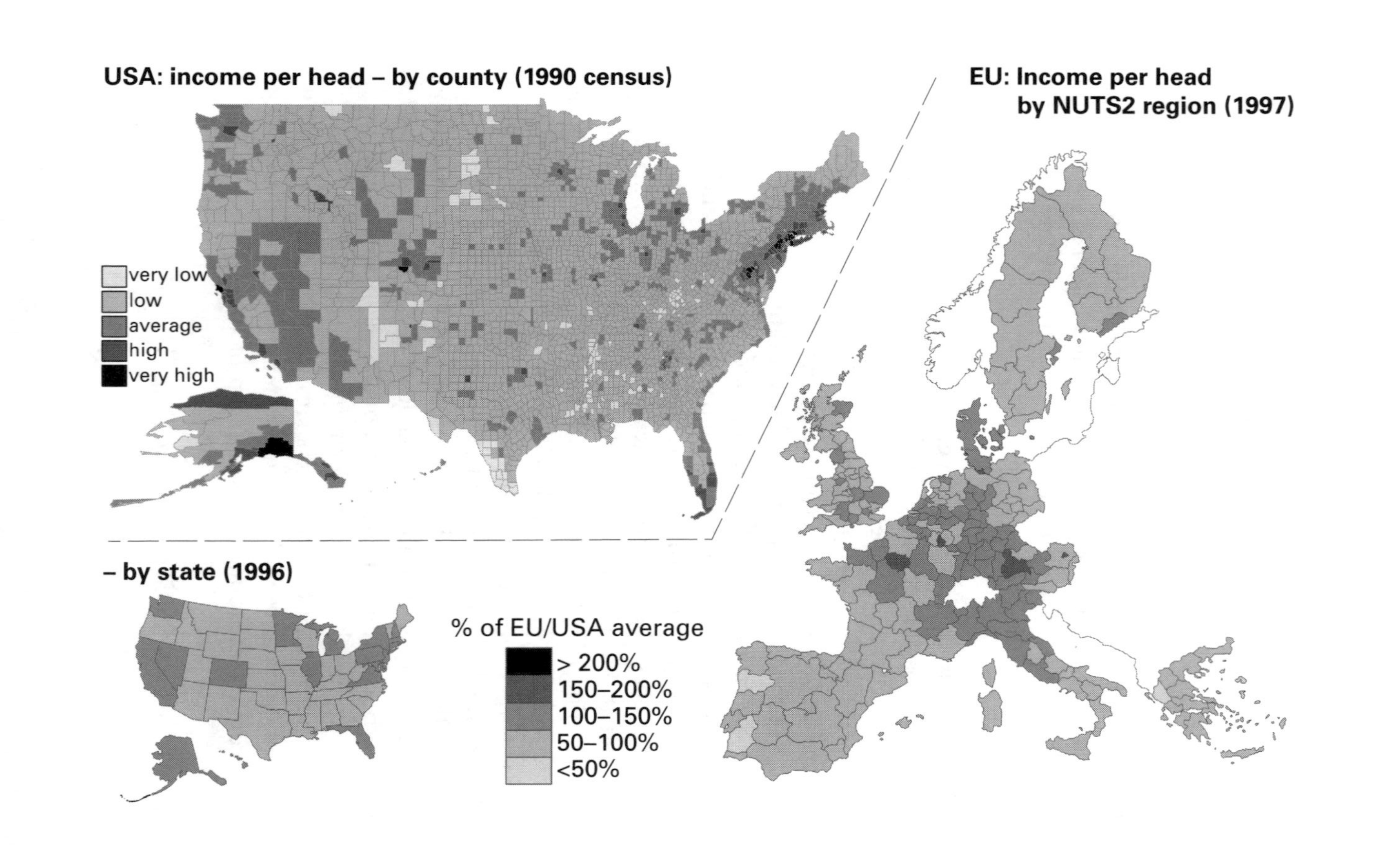
USA: income per head – by county (1990 census)
very low
low
average
high
very high
– by state (1996)
% of EU/USA average
> 200%
150–200%
100–150%
50–100%
<50%
EU: Income per head
by NUTS2 region (1997)

Regional differences in the USA and the EU

● The figures shown in the small map of the USA are for personal income per head as a percentage of the national average. For the EU the figures are regional product per head as a percentage of the EU average. The EU is divided into NUTS2 regions (nomenclature of territorial units for statistics). These correspond to German *Länder*, Spanish *autonomías* and French *provinces* and the equivalent in other EU countries. The large US map provides income data at the county level from the 1990 census.

○ The extent of regional variations depends on the size of the regions that are considered. In the USA the states have personal per-capita income levels ranging from $35,704 in Washington DC to $18,098 in Mississippi, a ratio of slightly less than 2 to 1. No state is more than 40 per cent above or less than 30 per cent below the national average.

The European data show more variation, but for smaller regions. Five regions (London, Paris, Hamburg and two more German regions) have per-capita levels more than 50 per cent above the EU average and twelve have levels less than 50 per cent of the average.

Despite the difference in the relative size of regions it is in fact almost certain that the range of difference in the EU is greater than in the USA since the European Union economy is still much less integrated than that of the USA.

But US regional variations are greater than they appear at the state level and something of this is conveyed by the map of five income levels divided by counties. It is possible to see the pockets of relative poverty in the Appalachian region, the states of the South and areas close to the Mexican border (›105).

■ REIS 1998; CIESIN 2000; Eurostat 1998.

Percent of families in income groups, 1992

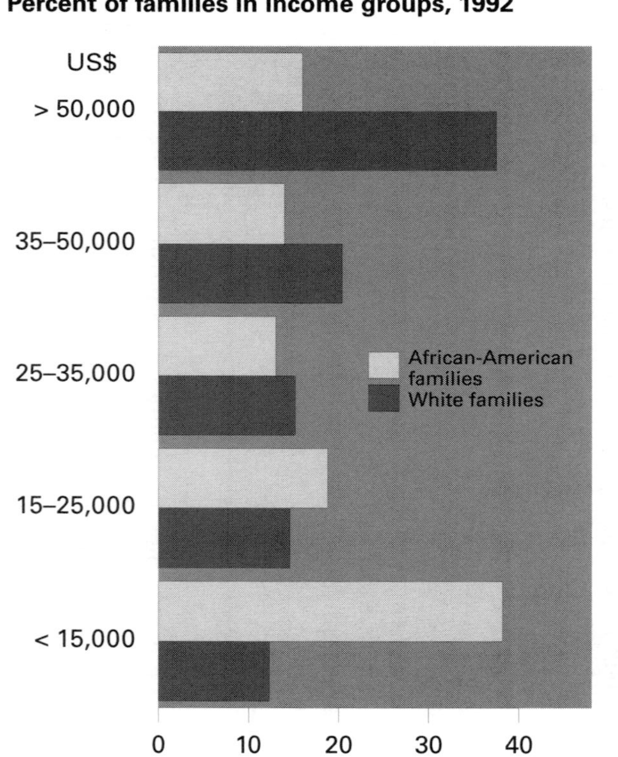

Blacks as multiple of whites, 1992

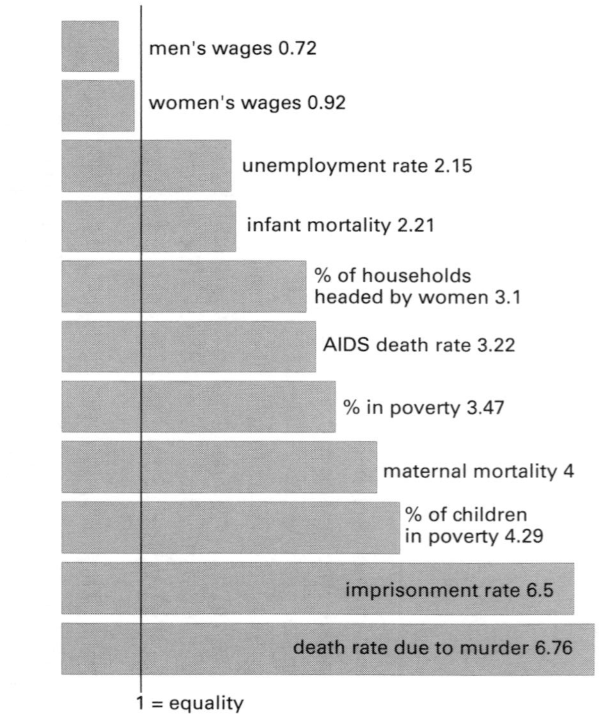

● The left-hand chart compares the percentage of white and African-American families in the USA in 1992 that fall into each of five income categories. And the right-hand chart shows several indicators for African-Americans as a ratio of the same indicator for whites in the USA.

○ The left-hand chart shows a striking difference in the distribution of income between whites and African-Americans in the USA. We find a larger proportion of white families than of African-American families in the highest three categories and a larger proportion of African-American families in the lowest two. Almost 40 per cent of African-American but only 12 per cent of white families receive the lowest of the five categories of income. And 40 per cent of white, but only 15 per cent of African-American families receive the highest. Hispanics are between whites and African-Americans, and people of Asian origin are higher than whites.

The right-hand chart shows a selection of the numerous ways in which the black population of the USA suffers discrimination and enjoys a lower level of welfare than whites. Economic indicators show that wages for African-Americans are lower (especially for men) than for whites; that more than three times the percentage of people, and more than four times the percentage of children, live in poverty; and that the black rate of unemployment is more than double the white rate. The health indicators show that African-Americans have double the rate of infant mortality, three times the AIDS death rate and four times the rate of maternal mortality of whites. Finally they have 6.5 times the incarceration rate (› 111, 112) and nearly seven times the death rate due to murder.

Hacker 1995.

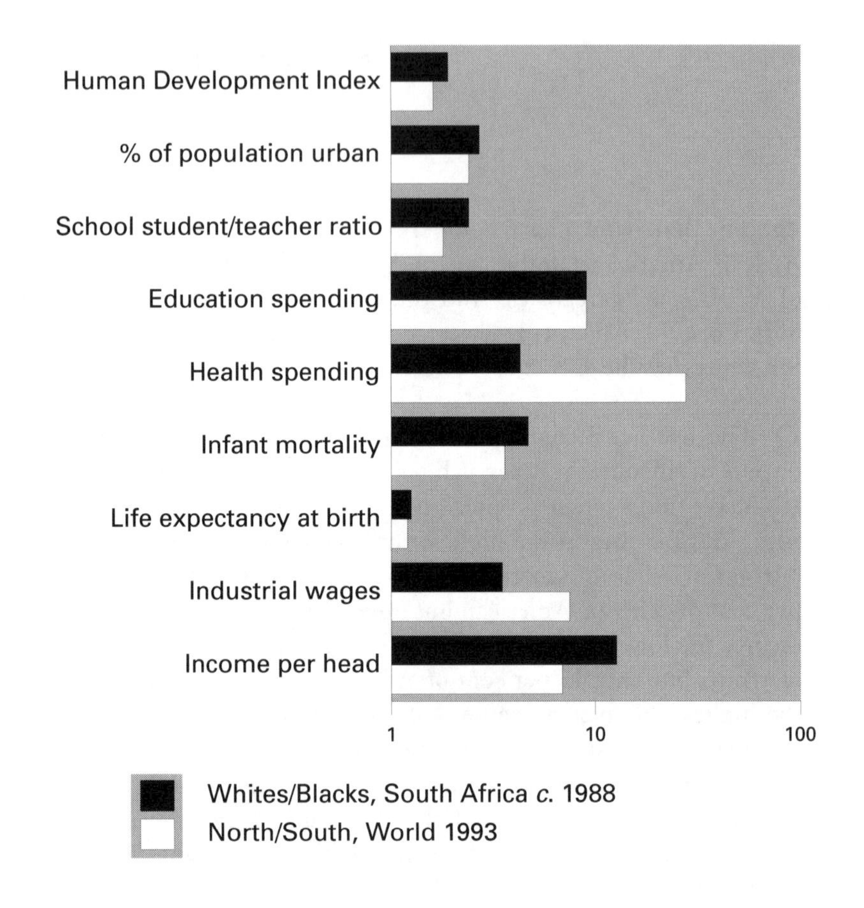

South Africa: the HDI disaggregated by social group, 1994

The world as a macrocosm of apartheid

● The graph shows a comparison of white/black differences in South Africa in the closing years of apartheid with differences in the world between North and South. The bars represent ratios of the values of a number of indicators for whites compared with blacks in South Africa (the black bar) and for the North compared with the South in the world (the white bar). Where the white bar is longer than the black one the North/South inequality in the world is greater than white/black inequality in South Africa and vice versa. The scale is logarithmic (progressively condensed) in order to show a very wide range of values.

○ Apartheid in South Africa was one of the most organized systems of discrimination the world has ever seen. It produced and maintained some of the inequalities that the black bars in this diagram display. The differences are underlined by the figures in the lower chart for the Human Development Index disaggregated according to race, and then sex and urban/rural residence. Whites in general had in 1994 an HDI of 0.878 while rural black women had a rate of only 0.356. It is the difference between Spain and Haiti.

The upper chart also shows (the white bars) the discrimination that exists between North and South in the world. The interest of the counterposition is that it shows that the degree of North/South inequality in the world is strikingly similar to white/black inequality under apartheid. In some respects, most notably health spending, the North/South inequality is greater. In other words, the world seems to be a macrocosm of a discriminatory system which that same world almost unanimously rejected as morally unacceptable.

■ UNDP 1964; UNDP 1966; SAIRR 1990; Wood 1994.

The international economy

VI

International trade,
foreign investment,
international institutions,
the external debt
and development aid

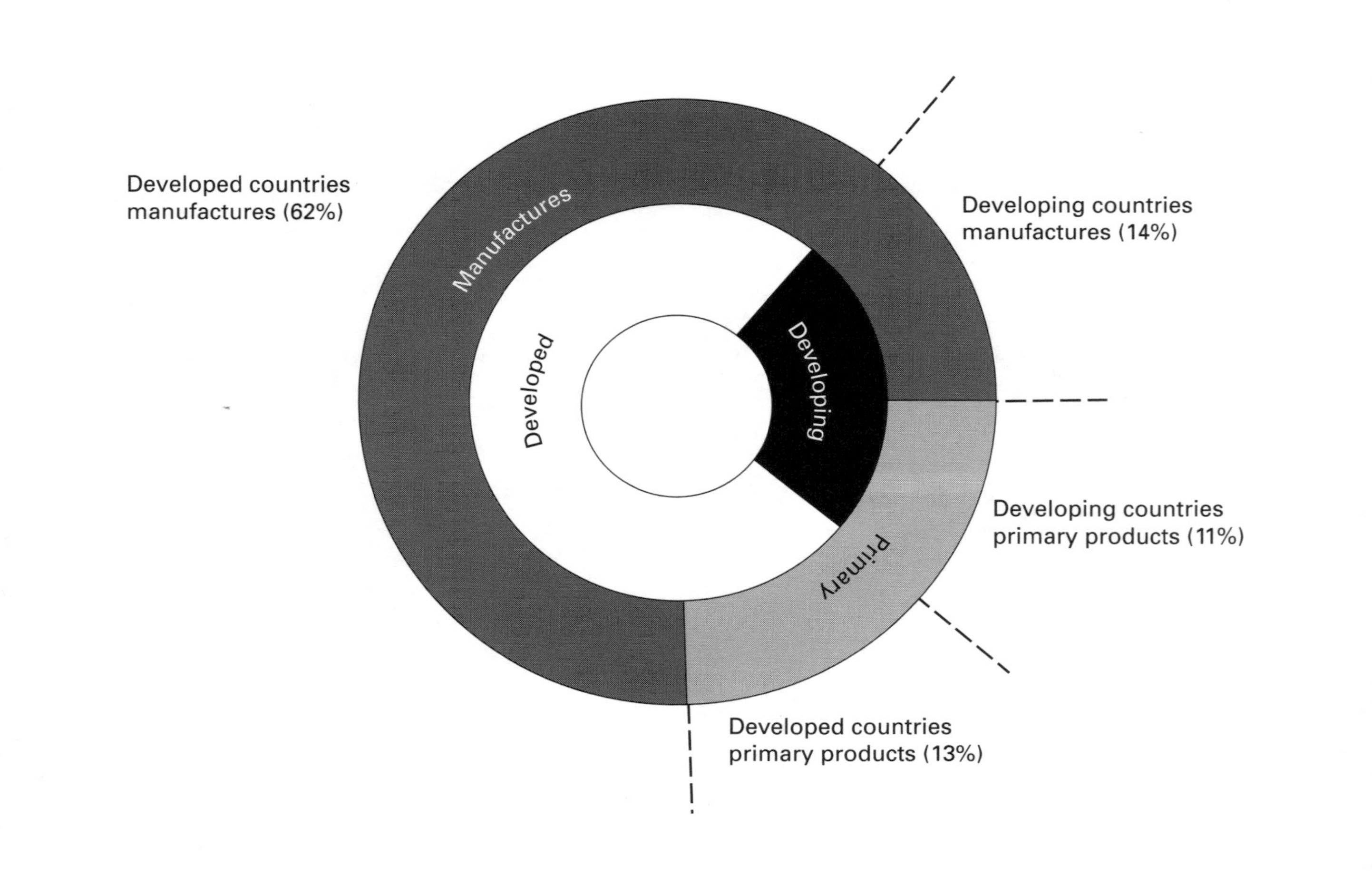
Developed countries manufactures (62%)
Manufactures
Developed
Developing
Developing countries manufactures (14%)
Developing countries primary products (11%)
Primary
Developed countries primary products (13%)

Who exports what? The structure of world trade

● This graph has the same form as graph 59. The outer ring shows the division of world exports between manufactures and primary products, and the inner ring shows their division between developed and developing countries. The two rings are then adjusted in relation to each other to show four categories of world trade: exports of manufactures and primary products from developed and developing countries.

○ The information in this chart will surprise the many who believe that developed countries export manufactures and developing countries export primary products. The structure of world exports can be summarized as follows:

- 75 per cent of the world's exports are from developed countries and only 25 per cent from developing ones;
- developed countries export mainly manufactured goods (83 per cent of their total, 62 per cent of all world exports);
- developing countries also export more manufactures than primary products (56 per cent of their total, 14 per cent of world exports);
- more primary products are exported by developed countries than by developing countries (14 per cent of world exports, compared with 11 per cent).

Some of these facts go very much against the grain of what is generally believed. The beliefs were once correct. The structure of world trade has in the last three decades been going through an important change.

■ UNCTAD 1999a.

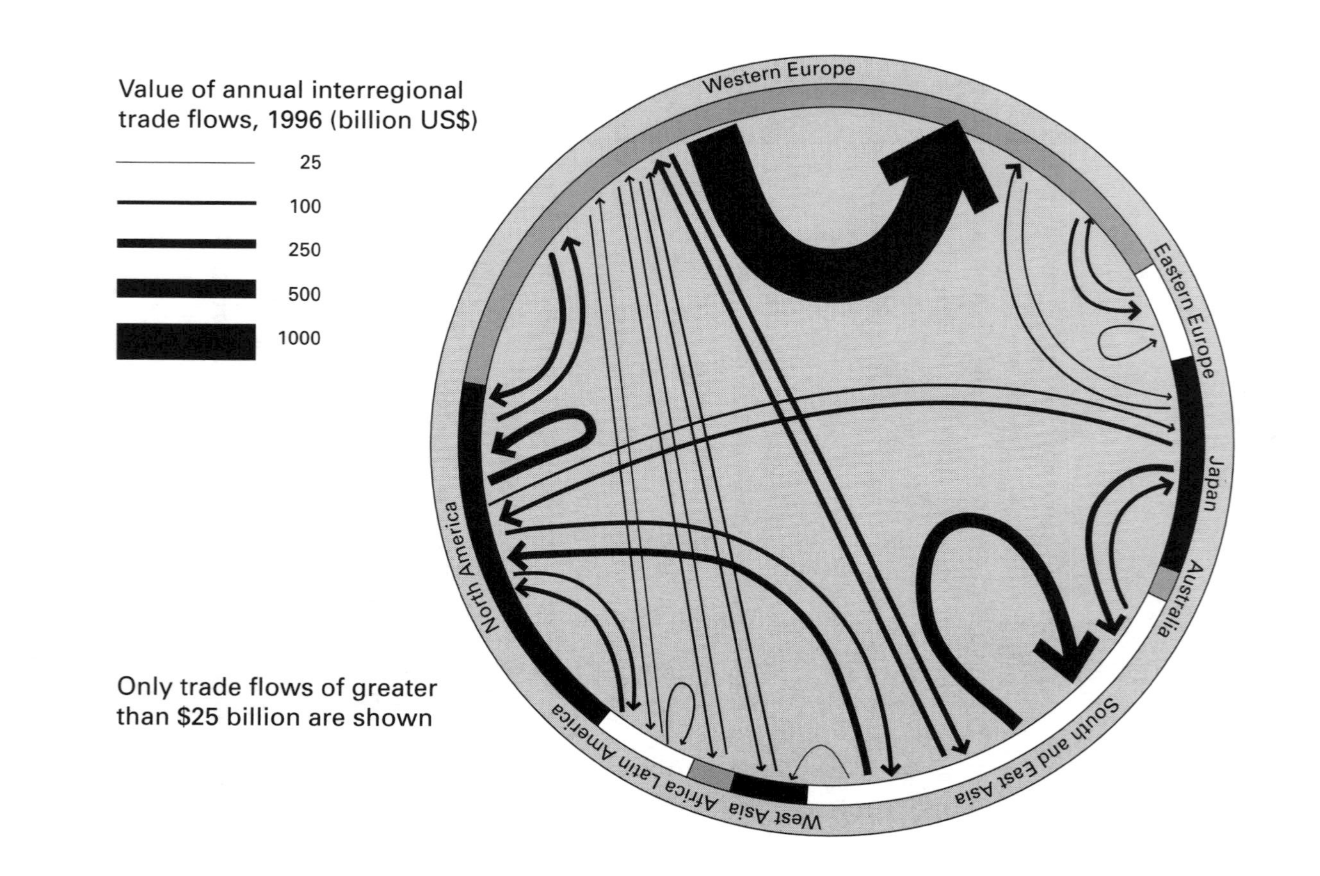
Value of annual interregional trade flows, 1996 (billion US$)
25
100
250
500
1000
Western Europe
Eastern Europe
Japan
Australia
South and East Asia
West Asia
Africa
Latin America
North America
Only trade flows of greater than $25 billion are shown

Who exports where? The direction of world trade

● This chart shows the general pattern of world trade between the major regions. The ring shows the share of the regions in total exports and the arrows show the amount and direction of the major trade flows (those of more than $25 billion annually).

○ International trade can be a very deceptive category. It seems to suggest a high degree of division of labour and specialization. But the fact is that to make a product in Vladivostok and sell it in St Petersburg, or to make a product in Reunion and sell it in France, is not counted as international trade although it exhibits a high degree of spatial division of labour. But to bake a tart in Hendaye and sell it in Irun is international trade, even though the distance between those two towns is a few metres. International trade means sales which cross juridical borders.

From this graph it can be seen that the largest trade flow in the world by far is trade among Western European countries. So, if the European Union became a single country, international trade would shrink by a large amount even though no change had occurred in the degree of division of labour in the world.

This graph shows that the three major trade flows in the world are within clearly defined regions (Western Europe, South and Southeast Asia and North America) and not between regions. It also shows the importance of the rich markets of Europe, North America and Japan for the exports of the rest of the world. And it draws attention to the very small role of Latin America and Africa in international trade flows.

■ IMF 1996.

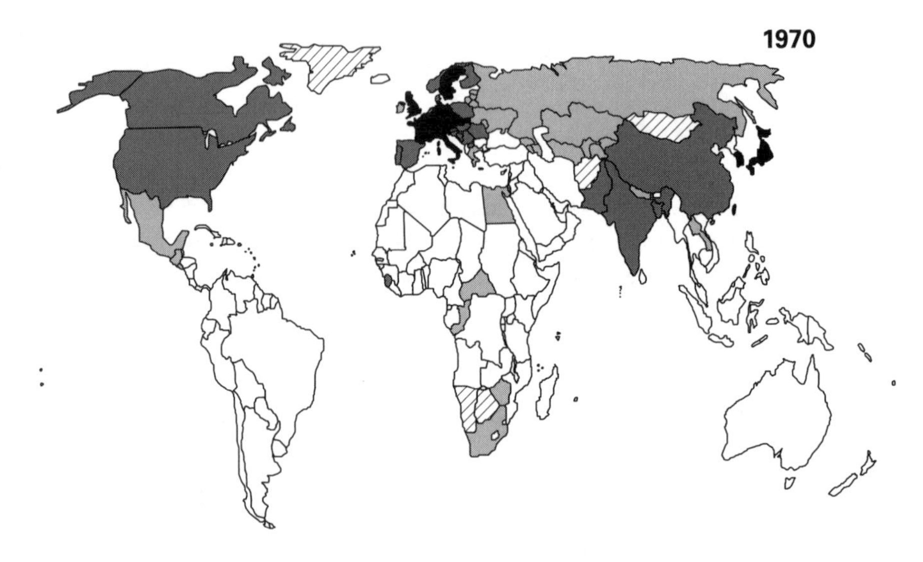

Manufactures as % of exports

no data 0–25 26–50 51–75 76–100

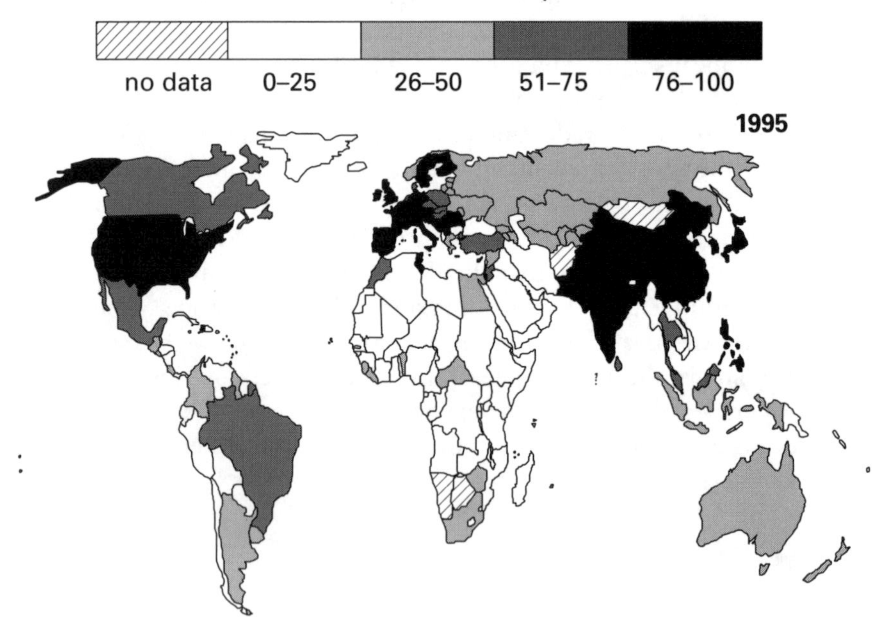

The growing industrialization of trade

● These two maps show the proportion of manufactures (as opposed to primary products) in exports of goods in 1970 and 1995.

○ The graph helps us to see the transformation that has taken place in world trade during the last quarter-century. As a proportion of total exports, primary products have fallen while manufactures have risen. The great majority of countries have been affected by this trend. Some countries which previously exported a majority of primary products now export more manufactures than primary products (for example, Ireland, Thailand, Malaysia, the Philippines, Brazil, Mexico, Morocco and Tunisia). Some countries which already exported more manufactures than primary products in 1975 have seen this balance move even more towards manufactures (for instance, the USA, China, Japan and Bangladesh). Others which still export a majority of primary products have seen the proportion of manufactures rise (for instance, Indonesia, Australia and Argentina).

But the graph also tells another side of the story. There is a group of countries – a few in Latin America and the majority in Africa – whose structure of exports has not shifted towards manufactures. They have, therefore, failed to enter the most rapidly growing markets. Their products have often both fallen relatively in price and had stagnant or declining markets. The result, as will be seen in succeeding graphs, is that their share of world trade has diminished.

■ UNCTAD 1999a.

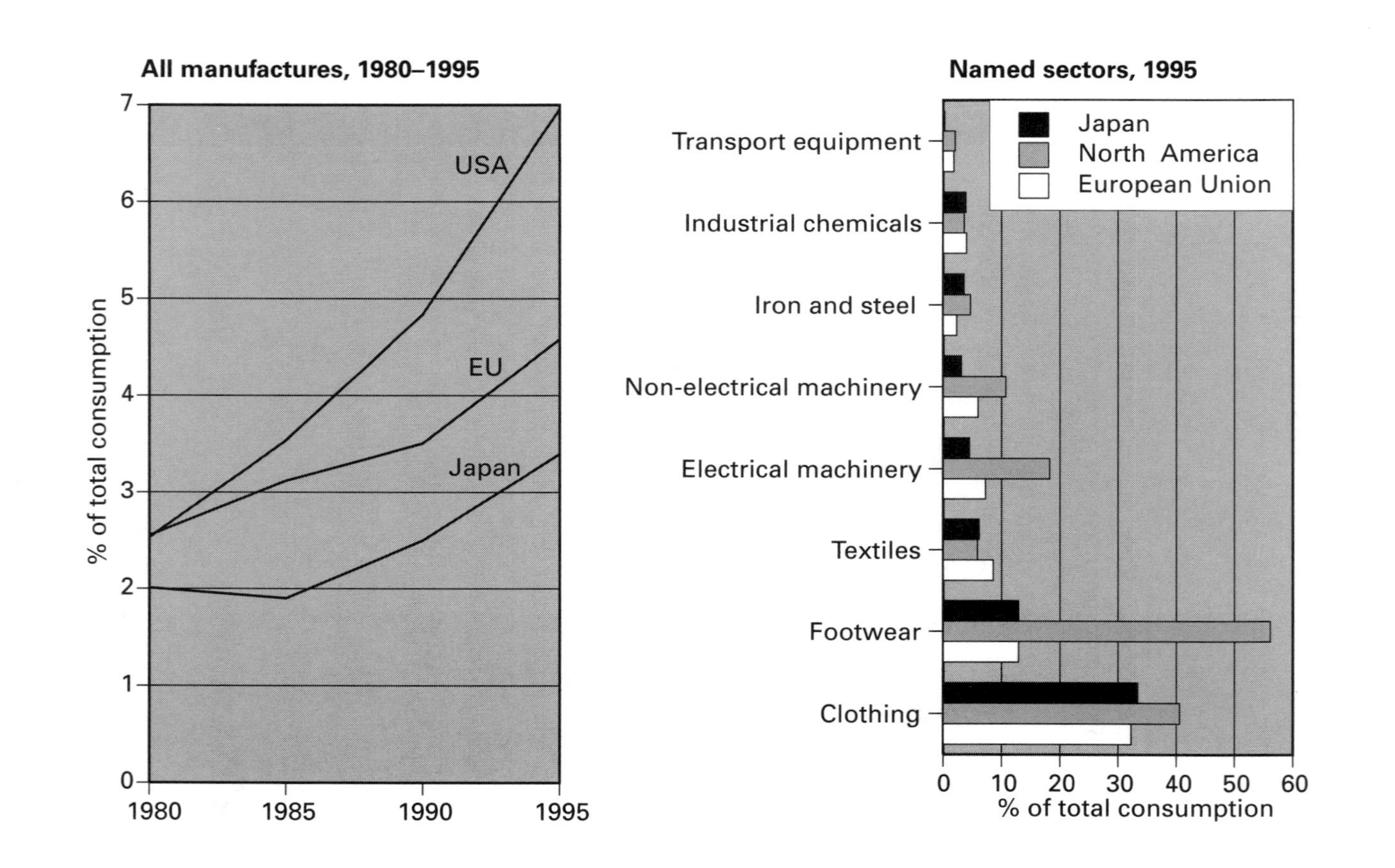
All manufactures, 1980–1995
% of total consumption
7
6
5
4
3
2
1
0
1980
1985
1990
1995
USA
EU
Japan
Named sectors, 1995
Japan
North America
European Union
Transport equipment
Industrial chemicals
Iron and steel
Non-electrical machinery
Electrical machinery
Textiles
Footwear
Clothing
0
10
20
30
40
50
60
% of total consumption

● The left-hand chart shows the value of manufactured imports from developing countries as a share of total consumption of manufactures in the three main developed-country markets between 1980 and 1995. The right-hand chart shows their share by selected sector for the same three markets in 1995.

○ These charts tell the same story of transformation in world trade from another angle – that of imports in the main consuming countries rather than that of exports. All the main developed countries have increased their imports of manufactured goods from developing countries during that last twenty years, as the left-hand chart shows. This process has gone furthest in the USA where in 1995 7 per cent of the consumption of manufactured goods was imported from a developing country. In Europe the figure was under 5 per cent and in Japan even less. The increases, especially to the USA, have made a huge difference to the prospects of manufacturing exports in poor countries. Nonetheless these percentages are much smaller than is normally believed by the public in the developed countries.

The reason for this exaggeration is seen in the right-hand chart. It is that the penetration of the market by developing-country producers has been especially large in the case of certain consumer goods like shoes and clothes and so is especially noticeable to the public. But in most other sectors, especially in the heavy industries and even in textiles, the penetration remains very small.

■ UNCTAD 1999a.

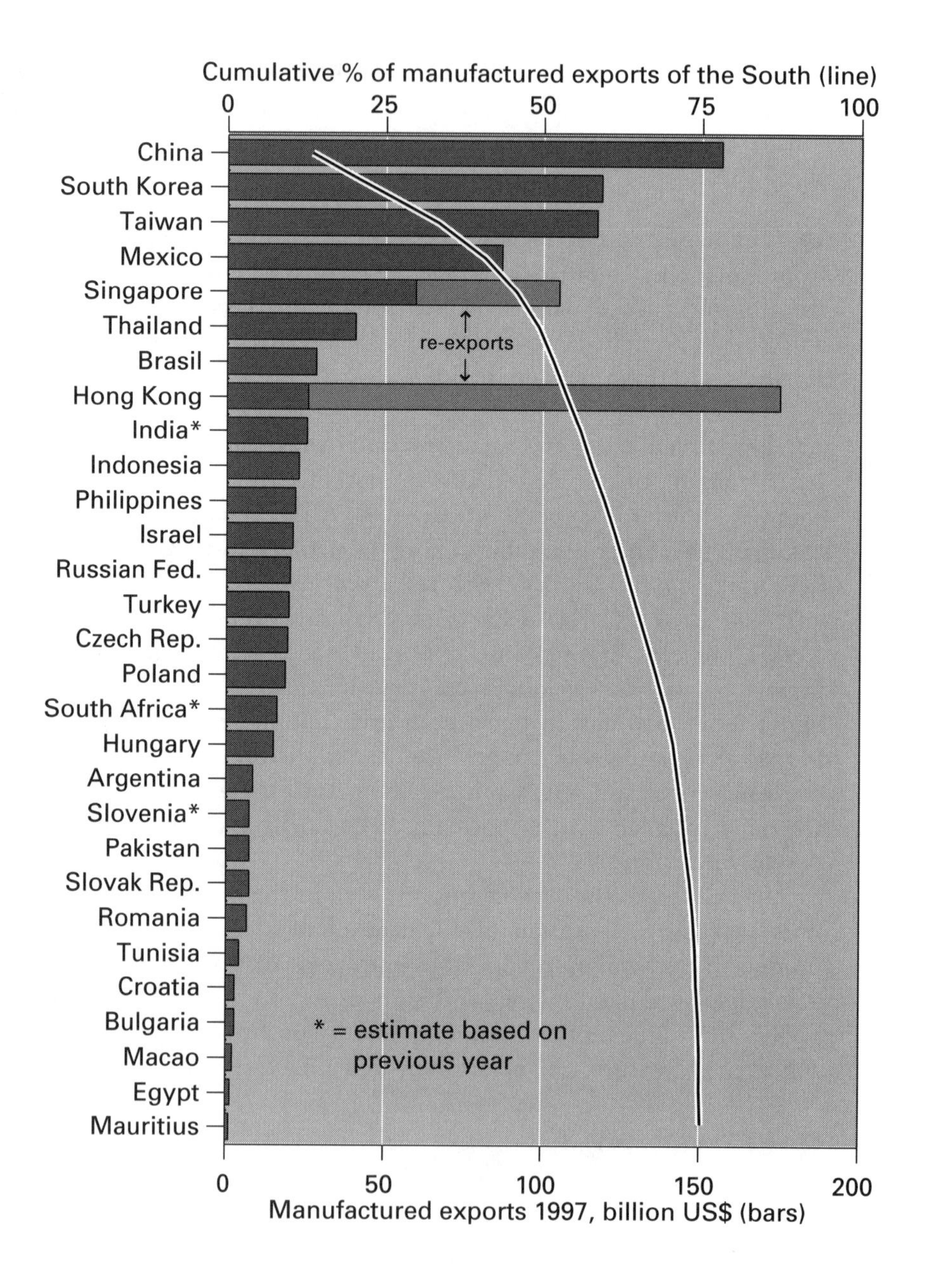

Cumulative % of manufactured exports of the South (line)
0
25
50
75
100
China
South Korea
Taiwan
Mexico
Singapore
Thailand
Brasil
Hong Kong
India*
Indonesia
Philippines
Israel
Russian Fed.
Turkey
Czech Rep.
Poland
South Africa*
Hungary
Argentina
Slovenia*
Pakistan
Slovak Rep.
Romania
Tunisia
Croatia
Bulgaria
Macao
Egypt
Mauritius
re-exports
* = estimate based on previous year
0
50
100
150
200
Manufactured exports 1997, billion US$ (bars)

● The bars on this graph (measured against the bottom axis) show the value of manufactured exports in 1997 from all those developing countries which exported more than $1 billion of manufactured goods. The line (measured against the top axis) shows the cumulative percentage which these countries account for of all the South's manufactured exports (China represents about 12.5 per cent, China plus South Korea nearly 25 per cent etc.).

○ This graph provides more detailed information about the new manufacturing exporters of the South. The largest is China, in spite of its still being subject to many impediments from its developed trading partners. The other really large exporters are South Korea, Taiwan and Mexico. Singapore and Hong Kong are large exporters but, as can be seen, a high proportion of what they export was previously imported from elsewhere. After that the amount of exports from each successive country falls quite fast. In other words, the development of exports from the South has been very unequally distributed between countries. These top twenty-nine manufacturing exporters account for (as can be read from the line indicating the cumulative percentage) three-quarters of all the South's manufacturing exports. This degree of concentration is what explains that, while the aggregate size and structure of the trade of developing countries has been transformed over the last thirty years, many countries have not participated at all in the process. Their exports remain as they were before world trade began to expand so quickly.

■ UNCTAD 1999a.

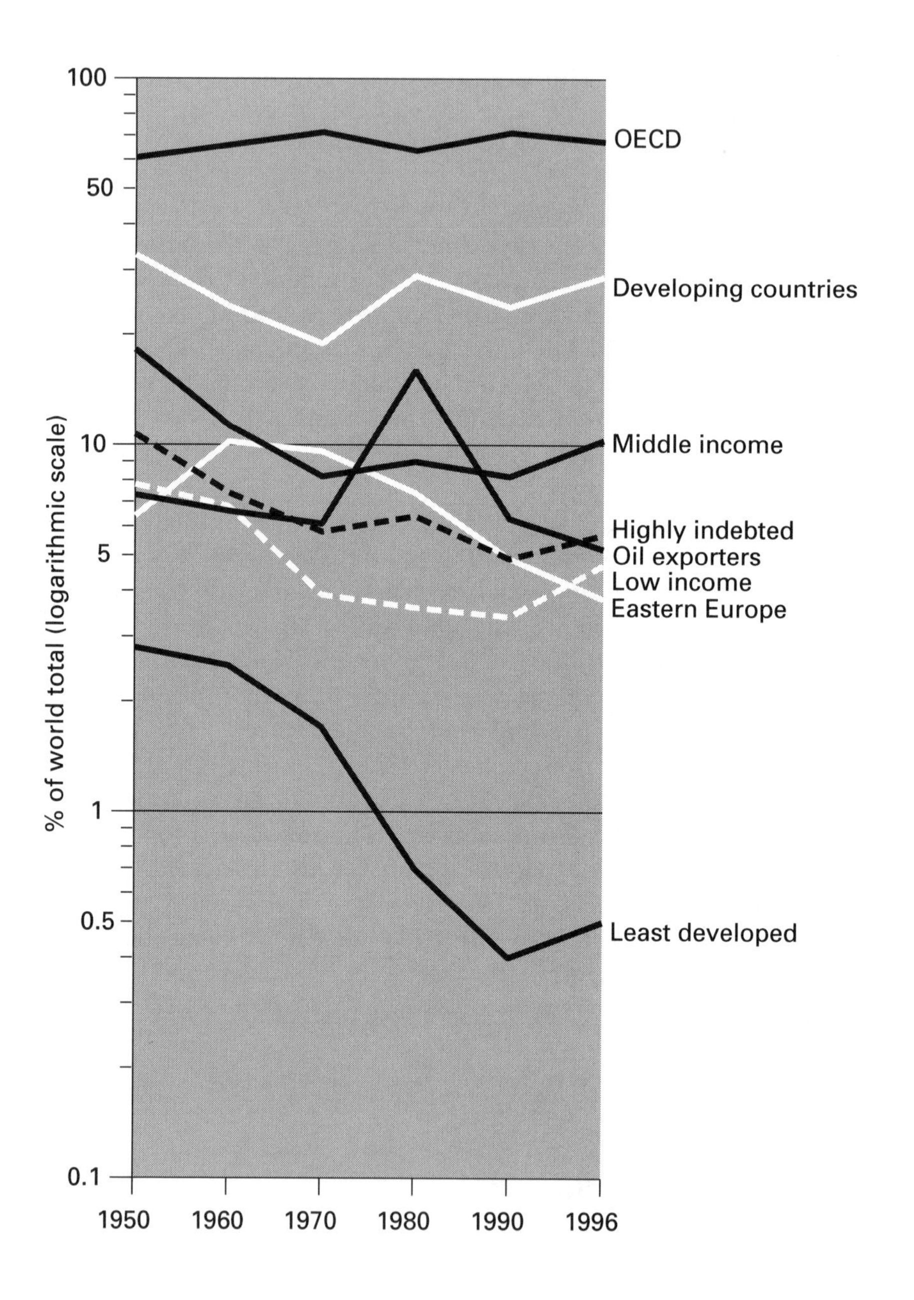
100
50
10
5
1
0.5
0.1
% of world total (logarithmic scale)
1950
1960
1970
1980
1990
1996
OECD
Developing countries
Middle income
Highly indebted
Oil exporters
Low income
Eastern Europe
Least developed

● This graph and the next show how world exports were divided between different groups of countries between 1950 and 1996. The scale is logarithmic so that large and small numbers can be seen together; as a result fluctuations in the smaller numbers are visually exaggerated compared with those of the larger numbers.

○ The evolution of the share of exports of a group of countries represents its participation in the world economy and so its access to purchasing power over imports. Export earnings have an important effect on both the real income of a country and its ability to develop. A declining share of exports will often mean, therefore, a declining share of world income. The top two lines show the division of total world exports between developed countries (**OECD**) and developing countries. From 1950 to 1970 the developed countries gained and the developing countries lost. There followed a decade in which this trend was reversed, largely due to the rise in the prices of petroleum (one of the three most important commodities in world trade) after 1973. In the 1980s the developing-country share fell back, largely due to the decline in petroleum prices, and finally in the 1990s the developing countries share increased as a result of the successes of new manufactures-exporting countries.

The lower lines divide countries into narrower groups. There is some overlap in the composition of the groups (between low income countries and least developed countries, for example). The rise and fall of the share of the oil-producing countries is very clear. In the 1990s all these categories of developing countries except the oil producers and the formerly communist eastern European countries increased their share of world exports, though only modestly.

■ UNCTAD 1999a.

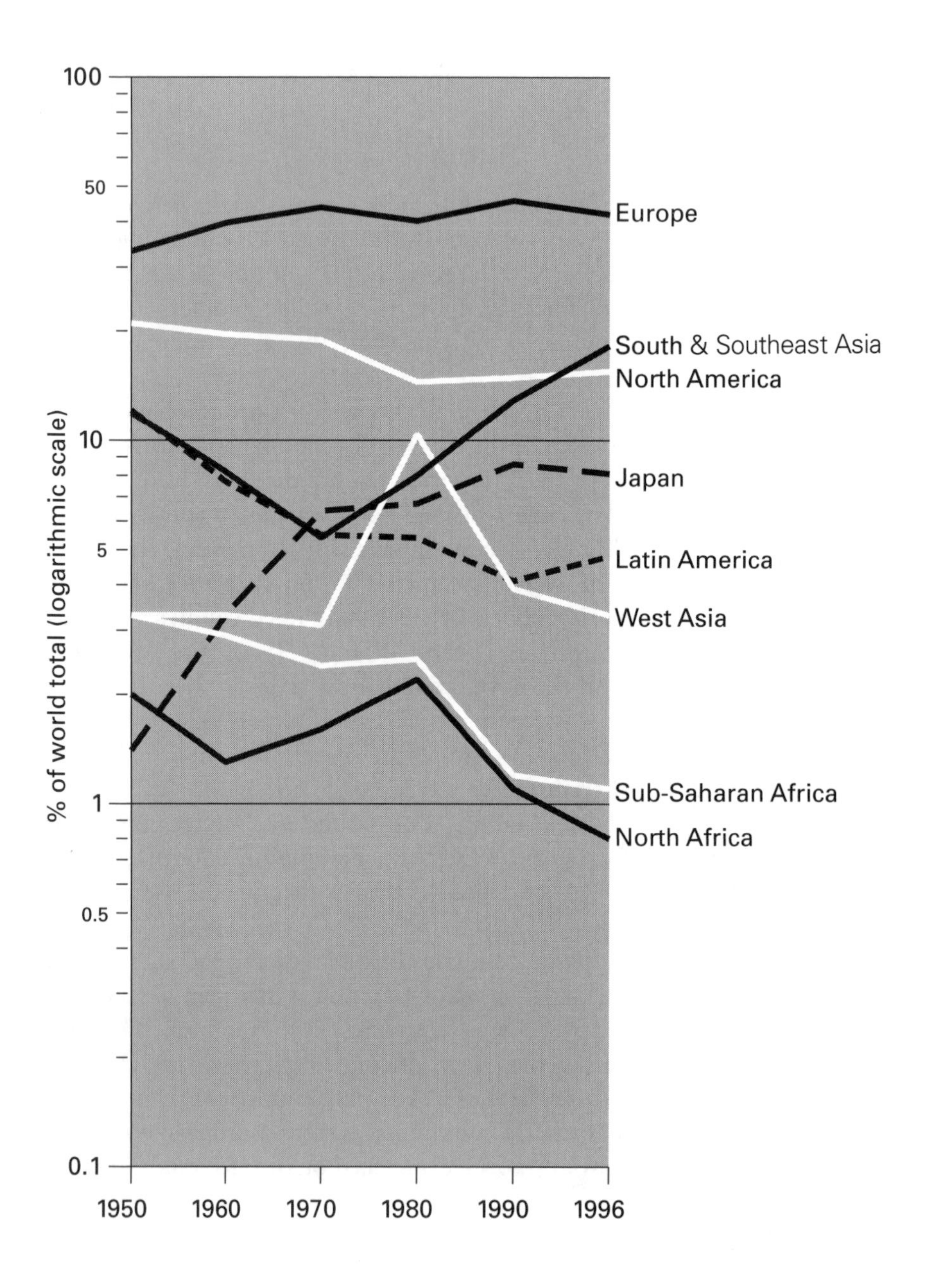

100
50
10
5
1
0.5
0.1
% of world total (logarithmic scale)
1950
1960
1970
1980
1990
1996
Europe
South & Southeast Asia
North America
Japan
Latin America
West Asia
Sub-Saharan Africa
North Africa

● This is identical to the previous graph except that countries are grouped according to geographical area and not analytical category.

○ While in the previous graph we saw that most of the categories of developing country increased their share of exports during the 1990s, here we see that this increase was in fact selective. Latin America's share increased a little, but after forty years during which it had fallen substantially. But the exports of West Asia and North Africa (both influenced by the oil price) and of Sub-Saharan Africa all fell. The rise in manufactured exports from the developing countries has come above all from the countries of South and Southeast Asia, which since 1970 have strongly increased their share of world trade.

Once more here is evidence of the special economic problems of Sub-Saharan Africa. There has been a historic decline in the exports of the continent. Its share of the world total has dropped from over 3 per cent in 1950 to barely 1 per cent in 1996. This has been largely due to the fact that Africa has not basically changed the products it exports and that the prices of these products have tended to fall. Hence, in exports and also, as we have seen elsewhere, in levels of income, the most striking polarization of recent decades has been between different parts of the countries called developing; and especially between East Asia and Sub-Saharan Africa. International trade is one of the mechanisms through which that polarization has taken place.

■ UNCTAD 1999a.

Export prices by type of commodity, 1960–1999 (1960 = 100)

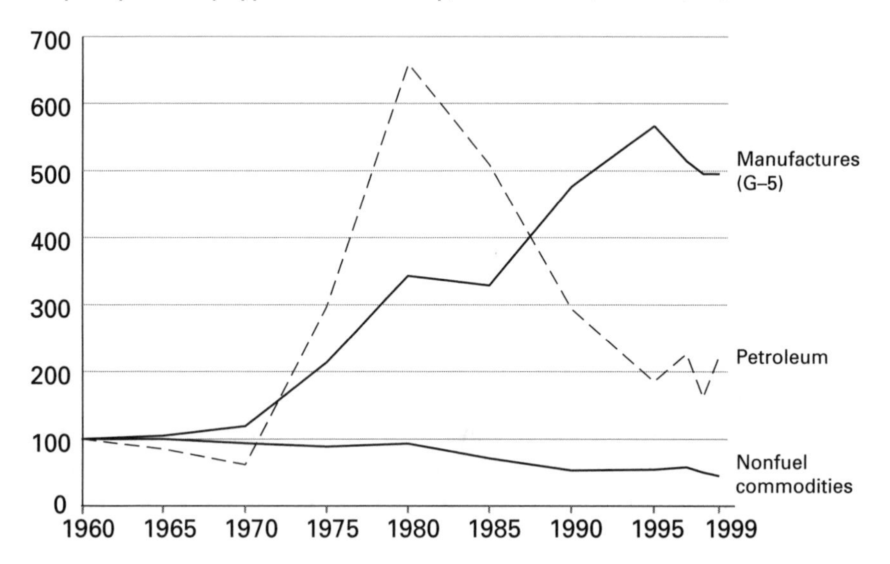

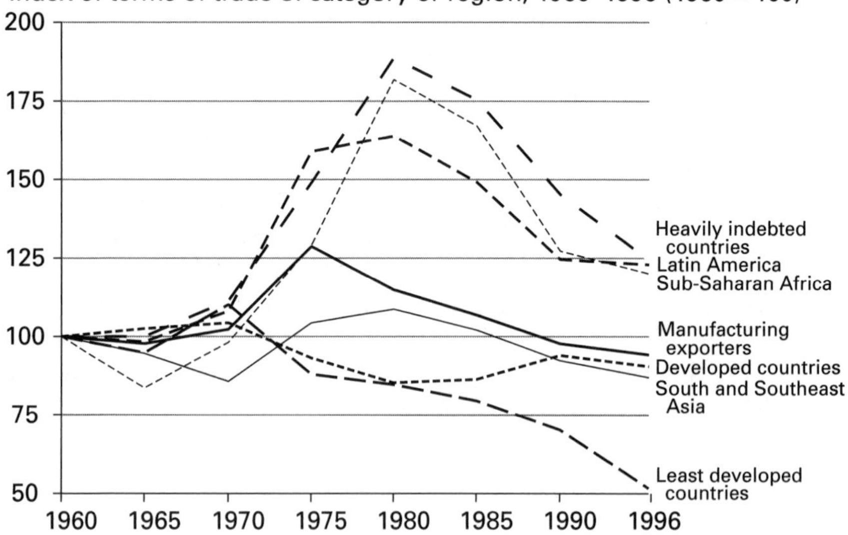

The problem of the terms of trade, 1960–1999

● The upper chart shows the evolution of world prices of three kinds of products since 1960. All prices are set at 100 in that year so that the graph compares the evolution of prices since then and not their level. The G5 countries are the 5 largest exporters of manufactures: USA, Germany, France, Japan and UK. The lower graph (also an index starting at 100 in 1960) measures the comparative evolution of the terms of trade for several regions and categories of countries. The terms of trade is the ratio between export and import prices; it is a measure of the real purchasing power of a unit of the country's exports.

○ While world commodity prices often shift up and down erratically, there is one long-term tendency: the prices of manufactures have tended to rise relatively and that of raw materials. So countries which export raw materials and import manufacturers tend to suffer adverse effects. They are obliged to sell increasing physical amounts of their export product in order to buy a given quantity of what they want to import. This is why, as we have seen in earlier graphs, producers of raw materials, especially those with limited demand and expanding supply like tropical beverages, suffer a particularly large fall in their national terms of trade. This reduces both their real income and their shares in world trade. The upper graph shows the relative price tendencies for manufactured and primary commodities; and the lower one shows the outcome for the terms of trade of different groups of countries. Since 1980 the terms of trade of the industrial countries have improved while those of the developing countries have worsened. The steady deterioration over four decades for the least developed countries is noteworthy.

■ World Bank 1999a; UNCTAD 1999a.

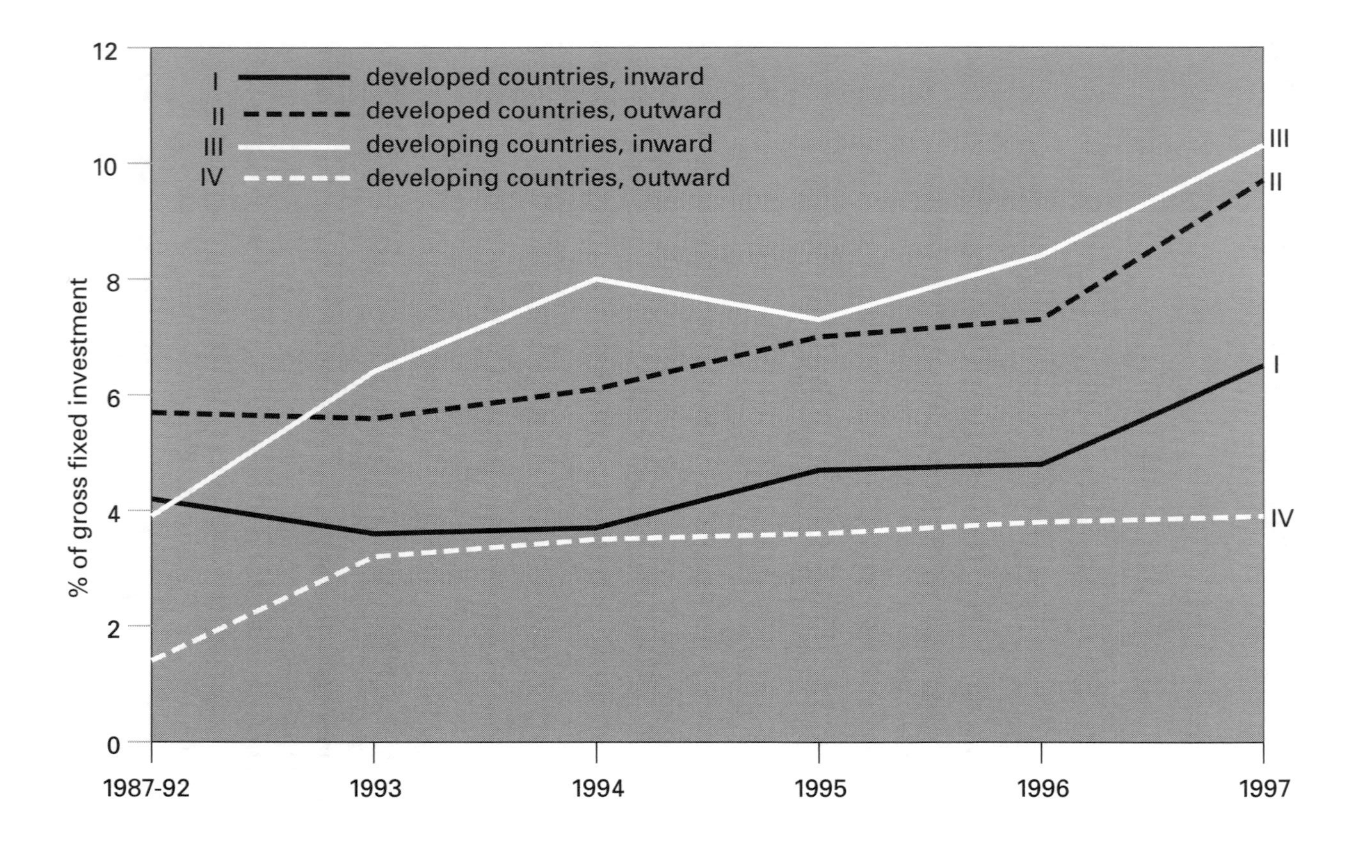
I developed countries, inward
II developed countries, outward
III developing countries, inward
IV developing countries, outward
% of gross fixed investment
12
10
8
6
4
2
0
1987-92
1993
1994
1995
1996
1997
III
II
I
IV

The growing importance of international investment

● The graph shows foreign investment as a percentage of total fixed investment. Investment here refers to direct investment (in productive assets) and not indirect investment (in financial assets). 'Inward' investment comes from other countries and 'outward' investment goes to other countries. The information is shown separately for developed and developing countries. In 1997 about 10 per cent of the fixed investment in developing countries came from abroad (line III) while developing countries invested in other countries a little under 4 per cent of the value of their fixed investment (line IV).

○ Here we turn from international trade to international capital flows. The graph shows the growth in recent years in the amount of investment that is foreign. Capitalists have invested a greater share of their total investments in other countries; and in many countries a higher share of investment is undertaken by foreign companies. Developed countries invest abroad relatively more than developing countries; and relatively more of investment in developing countries is from outside than in developed countries. But in absolute terms the great majority of foreign direct investment takes place between developed countries.

It is natural to interpret foreign direct investment as capitalists setting up new production facilities in other countries. In fact, however, the overwhelming majority of new investment activity in recent years has consisted of the acquisition of existing assets, often resulting from the privatization of state enterprises. Since privatization is a recent policy, and since it must logically have a limit (when all state assets are sold), it is possible that some of this rise in foreign investment will not be maintained in the long run.

■ UNCTAD 1999b.

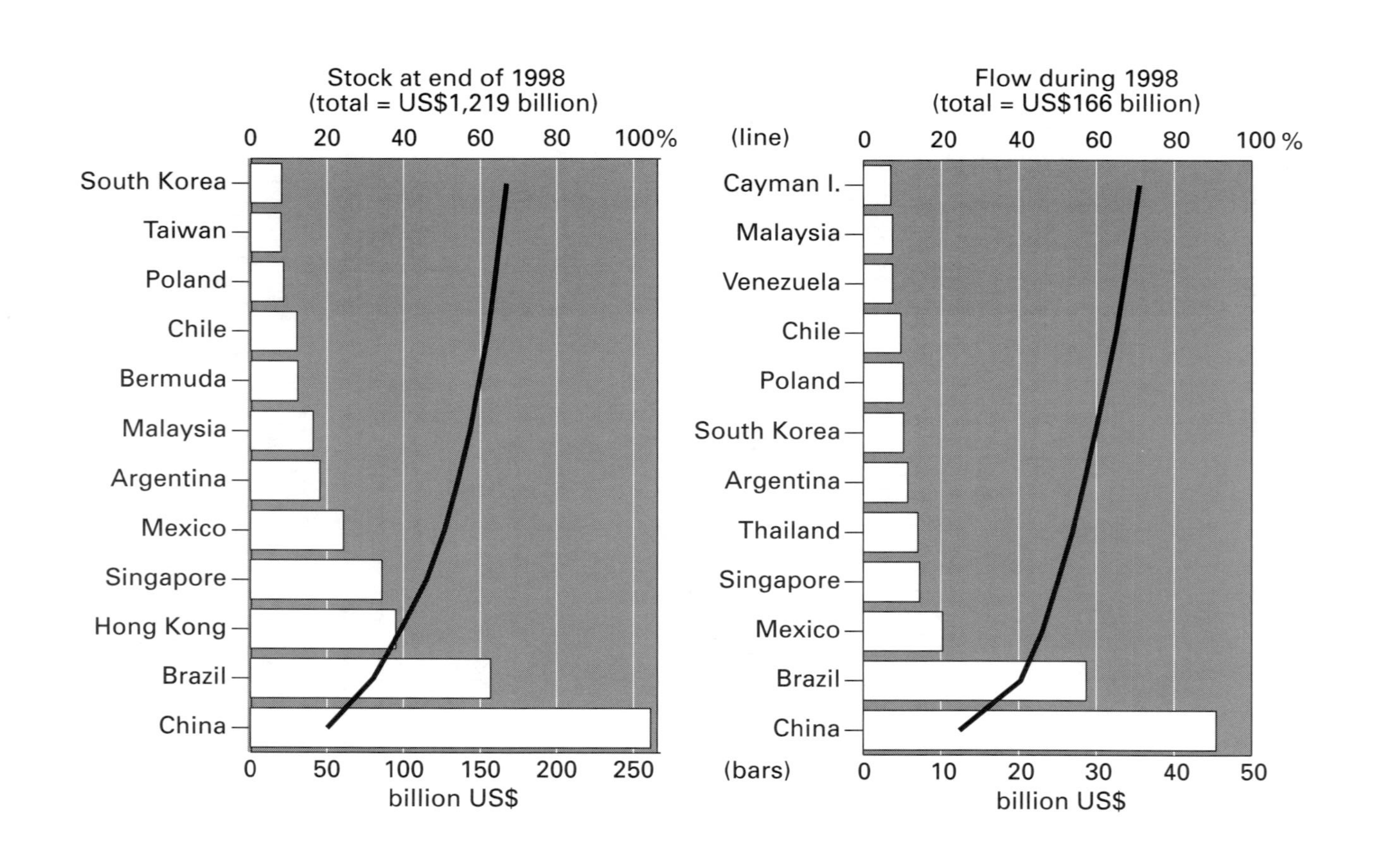
Stock at end of 1998
(total = US$1,219 billion)
0
20
40
60
80
100%
South Korea
Taiwan
Poland
Chile
Bermuda
Malaysia
Argentina
Mexico
Singapore
Hong Kong
Brazil
China
0
50
100
150
200
250
billion US$
Flow during 1998
(total = US$166 billion)
(line)
0
20
40
60
80
100 %
Cayman I.
Malaysia
Venezuela
Chile
Poland
South Korea
Argentina
Thailand
Singapore
Mexico
Brazil
China
(bars)
0
10
20
30
40
50
billion US$

Where the foreign investment in the South goes

● The two graphs use the same method to show the relative importance of different developing-country recipients of foreign investment. The left-hand chart measures the total stock of foreign investment in place and the right-hand chart measures the flow of new investments in 1998. The bars show the amount for the most important countries (measured against the bottom axis) and the lines (measured against the top axis) show (reading from the bottom up) the cumulative proportion represented by the countries on the chart of the total for all developing countries.

○ What this graph shows is that the flow of foreign direct investment into developing countries has been and remains extremely concentrated. Twelve countries account for about 70 per cent of both the stock and the recent flow of foreign capital in developing countries. The remaining 30 per cent is distributed among about 140 countries. Eight countries are in both lists, so the pattern is not changing much. By far the largest recipient country is China, followed by Brazil, Mexico and Singapore.

It is generally assumed that if developing countries import capital it must come from the developed countries of North America, Europe and Japan. But in the case of China, the most important developing country recipient of foreign investment, this is not so. The great majority of foreign investment in China comes from sources in Asia, particularly from the so-called Chinese Economic Area (Hong Kong, Taiwan and Macao) but also from countries such as Singapore, Malaysia and Indonesia where the 60 million overseas Chinese reside and control considerable amounts of capital.

■ UNCTAD 1999b.

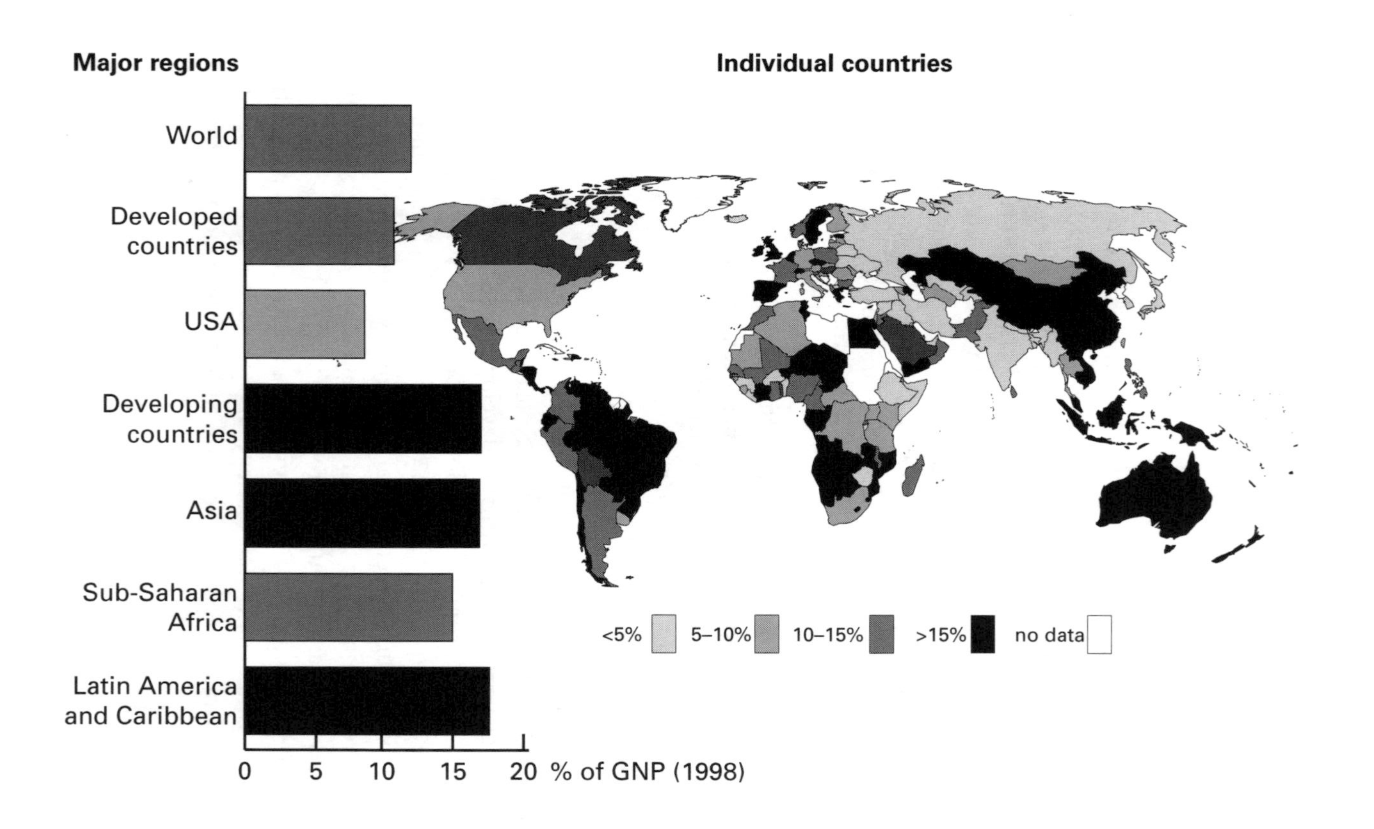
Major regions
Individual countries
World
Developed countries
USA
Developing countries
Asia
Sub-Saharan Africa
Latin America and Caribbean
0
5
10
15
20
% of GNP (1998)
<5%
5–10%
10–15%
>15%
no data

● The indicator shown in both the bar chart and the map is the stock of foreign capital in a country as a percentage of its annual national product. This is a hybrid measure used because there are no suitable data about the size of the capital stock. It gives a very rough idea of the relative importance of foreign capital in different countries.

○ It is difficult to detect much of a pattern in the importance of foreign capital in different countries. There are countries with a high and a low presence among virtually all groups and areas. It seems relatively more important in developing than developed countries, but the percentages are exaggerated because the national product figures used are converted using exchange rates and not **purchasing power parities** (‹ 8).

The country that contains by far the greatest amount of foreign direct investment is the USA, though it is low as a percentage of the national product. Among developed countries it is particularly important in the UK, Spain, Sweden, the Netherlands and Australia, and in the rest of the world in China, Brazil, Chile, Indonesia, Malaysia, Egypt and some West and Central African countries. The amount of foreign capital in Africa is slightly less than in Asia or Latin America, but the differences are not large.

These figures show that for the world as a whole foreign capital amounts to the equivalent of 12 per cent of the value of global production. That implies, as an extremely rough estimate, that about 5 or 6 per cent of the world's capital stock is owned by non-citizens of the country where it is located. This is probably much less than most people imagine. But it tends to concentrate in sectors of production which make it particularly visible while it is comparatively scarce in others.

■ UNCTAD 1999b.

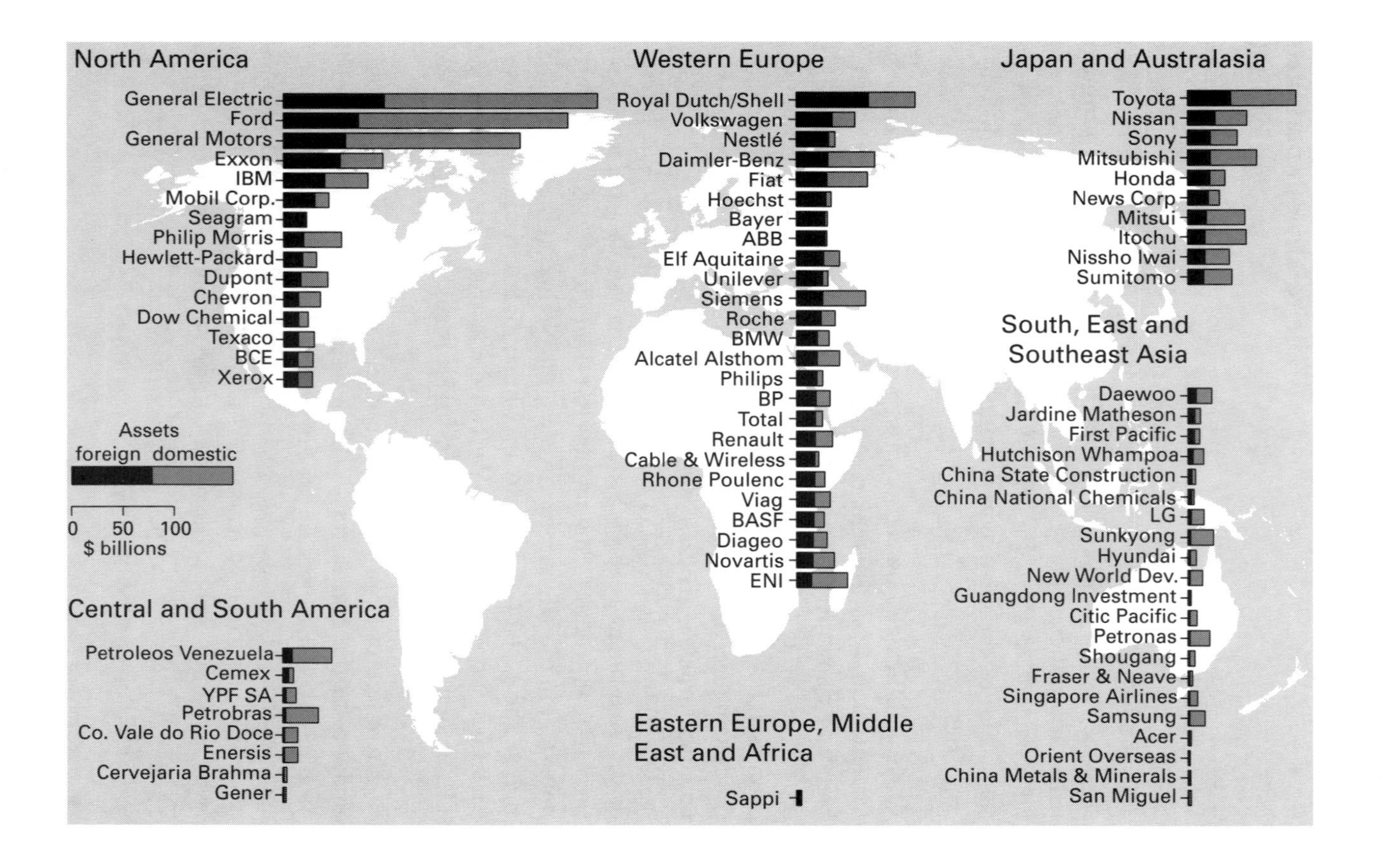
North America
General Electric
Ford
General Motors
Exxon
IBM
Mobil Corp.
Seagram
Philip Morris
Hewlett-Packard
Dupont
Chevron
Dow Chemical
Texaco
BCE
Xerox
Assets
foreign domestic
0 50 100
$ billions
Central and South America
Petroleos Venezuela
Cemex
YPF SA
Petrobras
Co. Vale do Rio Doce
Enersis
Cervejaria Brahma
Gener
Western Europe
Royal Dutch/Shell
Volkswagen
Nestlé
Daimler-Benz
Fiat
Hoechst
Bayer
ABB
Elf Aquitaine
Unilever
Siemens
Roche
BMW
Alcatel Alsthom
Philips
BP
Total
Renault
Cable & Wireless
Rhone Poulenc
Viag
BASF
Diageo
Novartis
ENI
Eastern Europe, Middle East and Africa
Sappi
Japan and Australasia
Toyota
Nissan
Sony
Mitsubishi
Honda
News Corp
Mitsui
Itochu
Nissho Iwai
Sumitomo
South, East and Southeast Asia
Daewoo
Jardine Matheson
First Pacific
Hutchison Whampoa
China State Construction
China National Chemicals
LG
Sunkyong
Hyundai
New World Dev.
Guangdong Investment
Citic Pacific
Petronas
Shougang
Fraser & Neave
Singapore Airlines
Samsung
Acer
Orient Overseas
China Metals & Minerals
San Miguel

Multinational corporations of the North and the South, 1999

● The bars show the value of assets, divided into domestic and foreign, of the fifty largest transnational companies of developed countries and the fifty largest of developing countries. They are divided in each case into three geographic blocks and ranked according to the value of their assets held outside their base country.

○ This graph provides a view of the major multinational corporations in the world. The very biggest of these firms, companies like General Electric, Ford and General Motors, hold most of their assets, and do most of their business, in their home countries. Others, companies like Seagram, Nestlé, Hoechst, Bayer, ABB and Unilever, do the great majority of their business abroad and are more truly multinational companies. Hardly any companies, however, have completely broken with a single national base, except long-established companies such as Unilever and Royal Dutch Shell. It remains to be seen if the recent trend in cross-border mergers produces more authentically international companies.

UNCTAD's official definition of a multinational company is a company that has at least one affiliate abroad of which it owns more than 10 per cent of the equity. According to this inclusive definition it is estimated that there are about 60,000 multinational parent companies with between them a total of about 500,000 foreign affiliates (about 8 per parent). These firms account for about 31 per cent of world production (23 per cent in the parent companies and 8 per cent in the affiliates). Virtually all large firms in the world are now in some sense multinational but many of the half-million affiliates are little more than sales agencies for the exports of the parent companies.

■ UNCTAD 1999b.

Voting rights

United States

Canada
Japan

Germany, France and UK

Other Europe

Africa, Latin America,
Caribbean, Asia and
Oceania

Who runs the IMF and the World Bank?

● The two pillars show the division of voting rights in the two major international organizations which regulate the world economy. These rights roughly correspond to the division of national incomes between countries.

○ As the world economy has become more integrated and as its inequalities and its potential instability have grown it has become clear to everyone that it requires institutions to control it, just as any national economy does. So far, however, the world's institutions are less powerful and harder to understand than national ones. In the first place a great deal of power and influence over the world economy is still exercised, as it has always been, by the economically dominant countries, especially the USA. But it is a hit-and-miss form of power, so they have tried to institutionalize it in various ways. There are informal groupings of countries which meet occasionally to try to coordinate policy. The main one of these is the Group of 7, now seemingly expanded to the Group of 8 to include Russia. It has so far been ineffective. There are two organizations which have no real power but a certain amount of global influence because of their access to and analysis of data: these are the Bank of International Settlements in Basle and the **OECD** in Paris. Membership of the latter is restricted to developed countries plus a few others for political expediency. The many organizations of the United Nations control a rather limited budget but produce a certain amount of influential analysis. But the two international institutions which have more money and more power, and occupy a unique situation between the rich and the poor countries are the two institutions established in the 1944 Bretton Woods agreement: the World Bank and the IMF.

■ World Bank 1999b; IMF 1999.

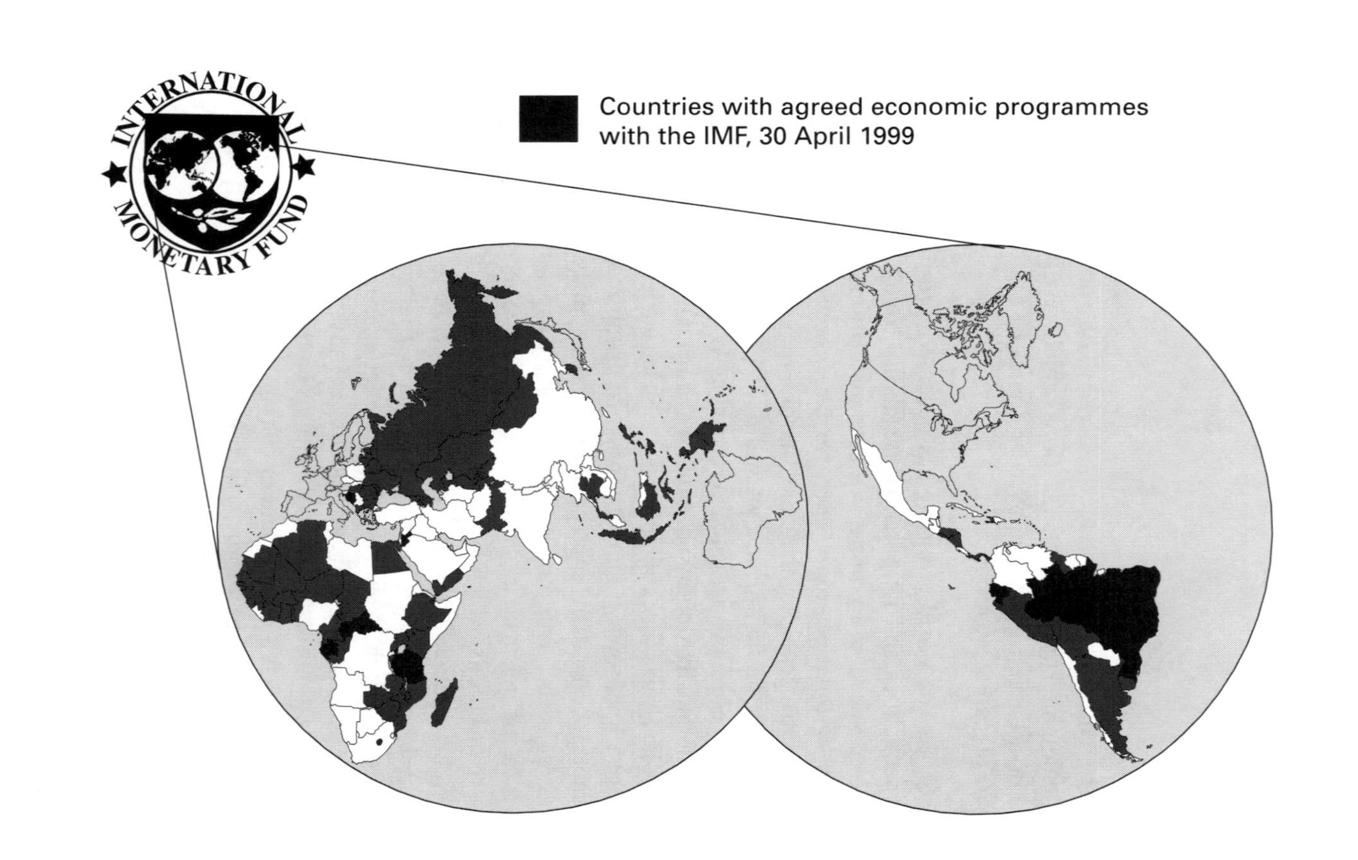
INTERNATIONAL MONETARY FUND
Countries with agreed economic programmes with the IMF, 30 April 1999

● The map marks in black the countries that in 1999 had entered into an agreement with the IMF through which they received loans or financial assistance in exchange for an agreed programme of economic policies.

○ In its brief history the IMF has lost one role, found another and may now be losing that too. Founded at the Bretton Woods conference of 1944, its job was seen as policing the new gold exchange system of fixed exchange rates. This system lasted more than twenty years. But between 1971 and 1973 it broke down, largely because the US administration decided to destroy it, by devaluing the dollar, breaking its link with gold and ending the fixed exchange rate system. With the end of fixed exchange rates the IMF had very little left to do, until it found a new role for itself after the debt crisis of 1982 (› 86, 87). It became a new financial police force in the indebted countries of the South, stepping in with loans and advice, on condition that troubled indebted countries adopted an approved economic programme. These programmes followed a pattern: privatization, deregulation and liberalization of both internal and external markets, the ending of state subsidies, budgetary balance and currency devaluation. This 'one size fits all' economic model became known as the structural adjustment programme. The map shows just how widespread these programmes had become by the late 1990s. The IMF's tentacles are now very widespread. Its reputation, however, has suffered a series of blows: widespread disillusion with structural adjustment programmes, failure to foresee the Asian economic crisis of the late 1990s, and the undignified and woefully undemocratic procedure by which its director, who had prematurely resigned, was replaced in 2000.

■ IMF 1999.

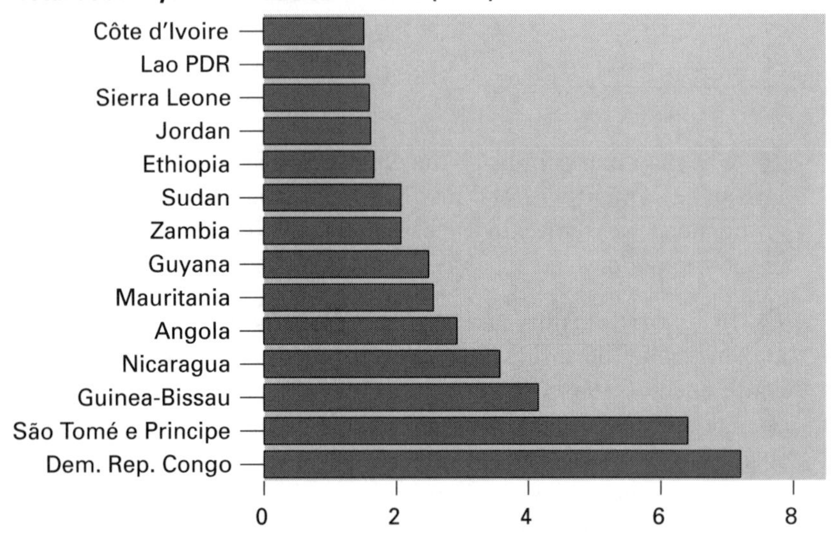
Total debt in years of national income (1998)
Côte d'Ivoire
Lao PDR
Sierra Leone
Jordan
Ethiopia
Sudan
Zambia
Guyana
Mauritania
Angola
Nicaragua
Guinea-Bissau
São Tomé e Principe
Dem. Rep. Congo
0
2
4
6
8

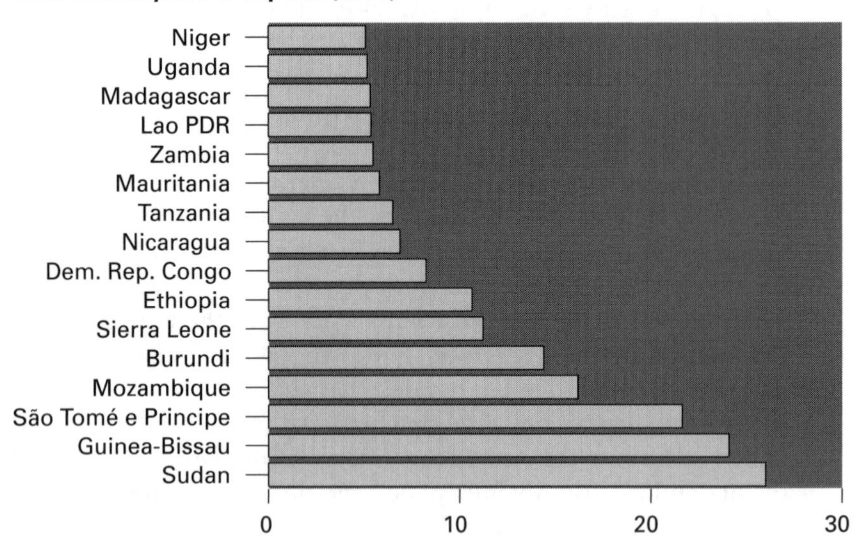
Total debt in years of exports (1998)
Niger
Uganda
Madagascar
Lao PDR
Zambia
Mauritania
Tanzania
Nicaragua
Dem. Rep. Congo
Ethiopia
Sierra Leone
Burundi
Mozambique
São Tomé e Principe
Guinea-Bissau
Sudan
0
10
20
30

● The bars show the countries which are the most indebted in relation to their national income. The debt is measured in multiples of their annual national income – in other words, in the number of years of their national income that they owe.

○ The most indebted country at the end of the twentieth century was the Democratic Republic of the Congo, which owed to its foreign creditors the equivalent of more than seven years of its annual national income. In this graph we see all the countries which owe two years' national income or more. This is a level of indebtedness similar to that which was imposed on Germany as reparations after World War I. John Maynard Keynes (in *The Economic Consequences of the Peace*) denounced that debt as unpayable and said it would be a source of future economic and political problems. For the countries shown here and more (mostly very poor African countries) the present debt is, as everyone knows, unpayable. A recent initiative to forgive some of the debts of the highly indebted poor countries (HIPC) has moved very slowly. Even if fully implemented it would still leave many of these countries highly indebted and it does nothing to eliminate the causes which drive them to borrow continuously in order to survive.

Since the debt has to be paid in foreign exchange it is perhaps even more telling to measure its burden by comparing it with a country's exports, as is done in the lower chart, which shows the countries that owe more than five years of the annual value of their exports. A number of countries are on both lists and in both of them Sub-Saharan African countries predominate.

■ World Bank 2000d.

Debt service as % of the national income (1998)

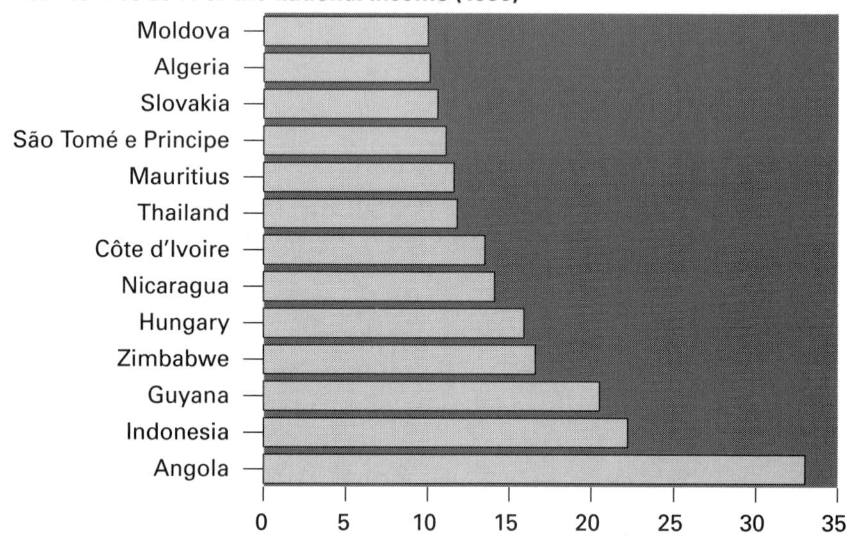

Debt service as % of exports (1998)

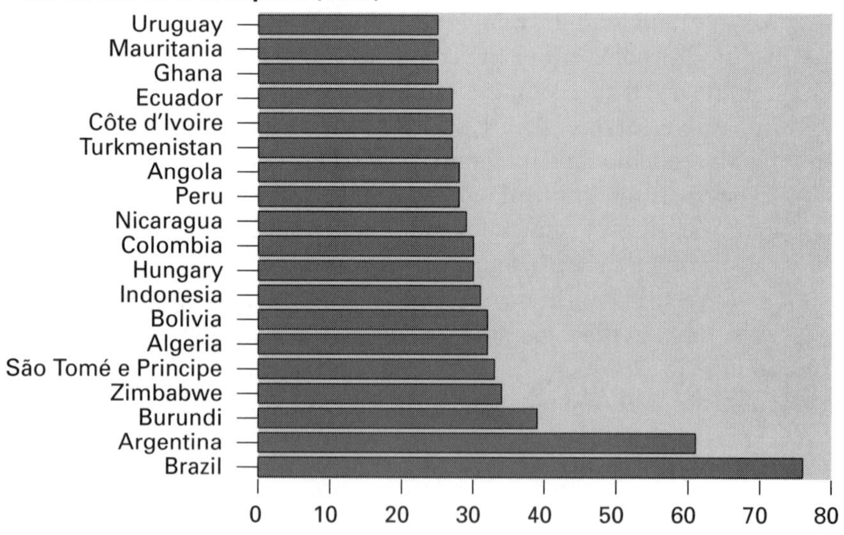

● The upper chart shows those countries which in 1998 paid more than 10 per cent of their national income in debt service (interest plus debt repayments). The lower chart shows those which paid more than 30 per cent of their exports in debt service over the years 1996 to 1998.

○ What countries actually pay in debt service is perhaps a better measure of the economic burden of the debt than an overall measure of indebtedness. A significant part of payments due are in fact not made. In 1999 arrears amounted to about $100 billion dollars of unpaid principal and nearly $50 billion of unpaid interest. Arrears may still contribute to the political burden of the debt since they permit the interference of creditors in the affairs of the debtors.

The figures in these charts, however, show the amounts actually paid in recent years by some of the major debtor countries in relation to both their national income and their exports. In the upper chart are the thirteen countries whose debt service amounted to more than a tenth of their national income in 1998. They are relatively diverse in both their geographical location and their level of income. To mobilize such sums, which do not result in any visible gain within a country, either puts a massive strain on a government's relations with its people or leads to a farcical process in which old debts are paid off with new debts. Neither is good for international stability or social progress. In the lower chart appear the nineteen countries which during 1998 had to use more than 30 per cent of what they earned by exporting to service their debts. Two major countries, Brazil and Argentina, were obliged to pay over 60 per cent of their export earnings for interest and debt repayment.

■ World Bank 2000d.

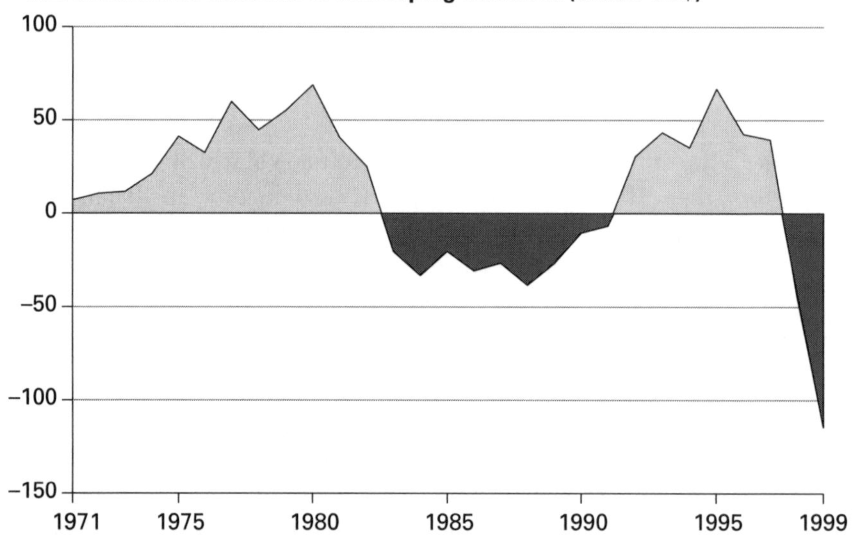
Net debt-related transfers to developing countries (billion US$)
100
50
0
–50
–100
–150
1971
1975
1980
1985
1990
1995
1999

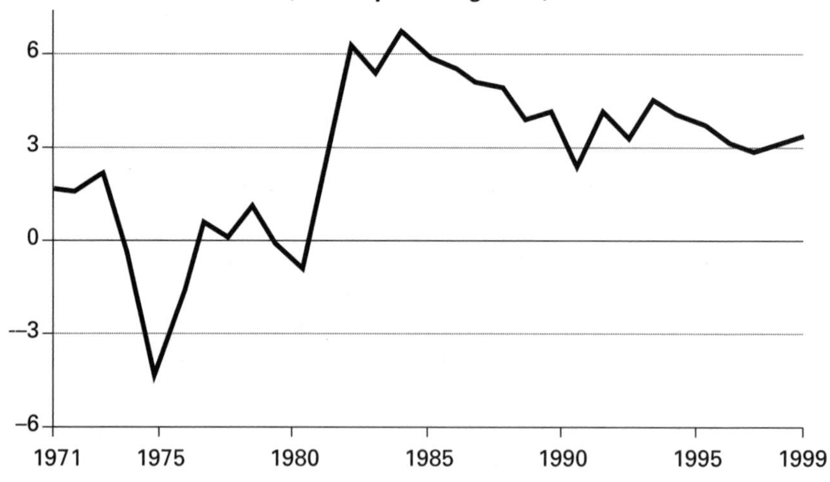
World real rate of interest (annual percentage rate)
6
3
0
–3
-6
1971
1975
1980
1985
1990
1995
1999

Some causes and effects of the debt crisis

● The upper chart shows the annual movement of debt-related transfers to and from the developing countries from 1971 to 1999. This figure is composed of new loans issued minus old loans repaid and interest paid. The lower chart shows for the same years the level of real world interest rates (the nominal rate of interest minus the rate of inflation), a measure of the true cost of borrowing.

○ Debt-related transfers are a measure of the amounts of money which actually flow to and from the indebted countries. The upper chart shows how developing countries borrowed money voraciously during the 1970s in response to the excess of funds available from the banks (due partly to the higher earnings of the oil-producing countries after OPEC raised the price of oil in 1973). Borrowing seemed desirable and easy because the cost of borrowing (the real rate of interest) was low, as can be seen in the lower chart. But after 1980 the austerity policies of governments such as those of Reagan in the USA and Thatcher in Britain caused inflation to fall, real interest rates to rise, raw material prices to fall and the value of the dollar, in which most loans were denominated, to rise. The borrowers found that they could no longer service their debts. They were prevented from defaulting by a series of renegotiations of the debt but the debt was now much more expensive and a large net flow of funds back to the lenders resulted. This lasted for a decade until a new positive flow began in the early 1990s. But the new flow of lending was more unstable and gave way to huge financial crises and a new record reverse flow of funds at the end of the century. In 1999, as a result of all aspects of the debt, over $100 billion flowed from the poor to the rich countries of the world.

■ World Bank 1996; World Bank 2000d.

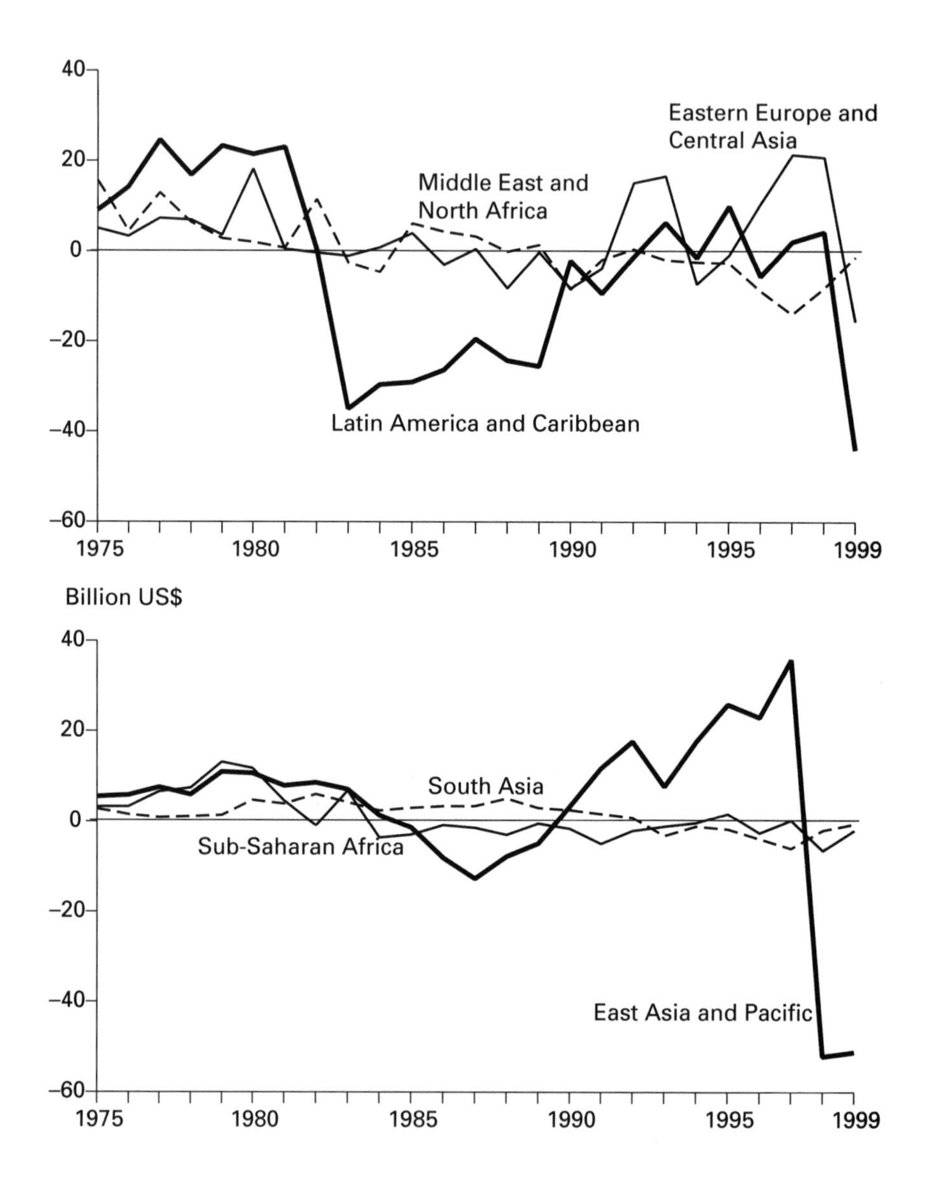

40
20
0
−20
−40
−60
Eastern Europe and
Central Asia
Middle East and
North Africa
Latin America and Caribbean
1975
1980
1985
1990
1995
1999
Billion US$
40
20
0
−20
−40
−60
South Asia
Sub-Saharan Africa
East Asia and Pacific
1975
1980
1985
1990
1995
1999

Positive and negative transfers: South to North aid

● The two charts on this page show the same information as the upper chart on the previous page, disaggregated into the six main areas for which data are available.

○ Here it is possible to see that the countries which were most adversely affected by the debt crisis of the 1980s were in Latin America and to a lesser extent in East Asia. During the 1980s more money related to the debt flowed out of Latin America than had flowed in during the 1970s, and Latin America has never regained a positive balance except in two isolated years. The crises at the end of the 1990s led to another enormous net outflow of funds.

East Asia was a large net receiver of funds after the end of the 1980s but the flow crashed even more severely than for Latin America at the end of the decade. Eastern Europe and Central Asia (the former USSR and COMECON countries) received two brief peaks of positive inflows during the 1990s but shared in the reversal of the end of the decade. For the Middle East and North Africa the net flow has got fairly steadily worse for three decades.

The transfer situation in South Asia and Sub-Saharan Africa has not been so volatile but has suffered a slow steady deterioration over two decades. The overall problem of the debt, therefore, despite innumerable plans to reform it, seems to be getting worse. 1999 was in fact the first year in which the net transfers related to the debt were negative for every one of the six areas. A system which in aggregate is responsible for growing net transfers from the poorer to the richer parts of the world is hardly likely, politically or economically, to add to international stability, not to mention justice.

■ World Bank 1996; World Bank 2000d.

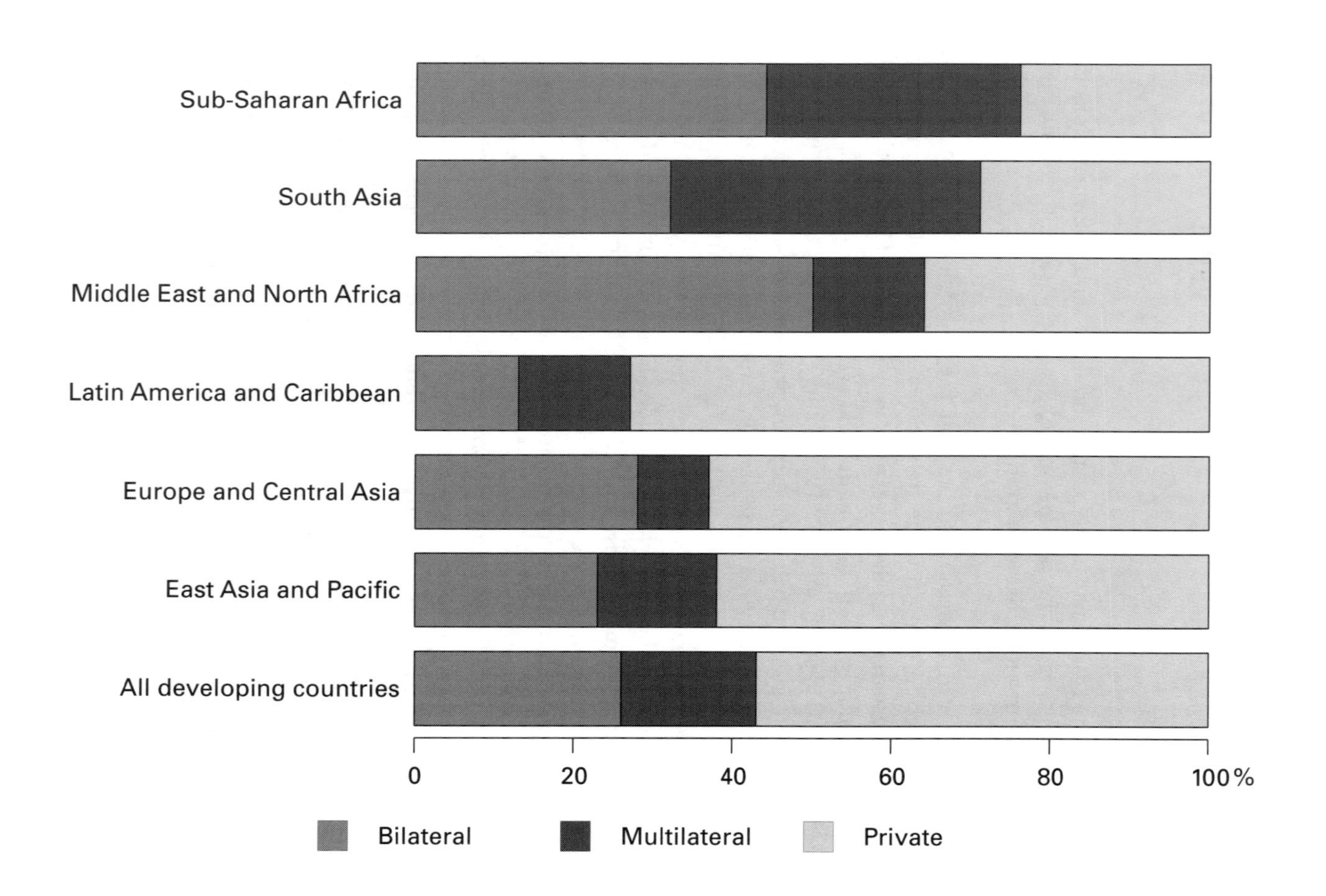
Sub-Saharan Africa
South Asia
Middle East and North Africa
Latin America and Caribbean
Europe and Central Asia
East Asia and Pacific
All developing countries
0
20
40
60
80
100%
Bilateral
Multilateral
Private

● The bars for the six regions and for all developing countries show the percentage of the long-term external debt owed in 1999 to three kinds of creditors: bilateral (other governments); multilateral (international financial institutions); and the private sector (banks, bondholders and firms).

○ The external debt of developing countries is far from being a homogeneous entity. It is owed on varying terms and conditions to an enormous variety of creditors. This chart shows the contrast between areas of the world in the composition of the debt with regard to different types of creditors.

A total of 57 per cent of the debt is now to the private sector, 17 per cent to the multilateral institutions and 26 per cent is bilateral debt to other states. But that division varies a great deal between different regions.

The poorest regions (Sub-Saharan Africa and South Asia) have the smallest proportion of private debt. Banks are not attracted by poverty. For countries in these regions, therefore, managing the debt is a question of international relations with the governments of rich countries and with the multilateral bodies. Creditor governments are often not too insistent on keeping up with payments, but they may use the debt, including slowly fulfilled promises to condone part of it, as a form of political leverage. The multilateral institutions are much less financially flexible. They only lend money on the strict understanding that they will be the first to be paid what is due. They have very little latitude to forgive debt unless this is financed by the richer countries. In the regions where private creditors predominate, managing the debt is more a part of general financial management.

■ World Bank 2000d.

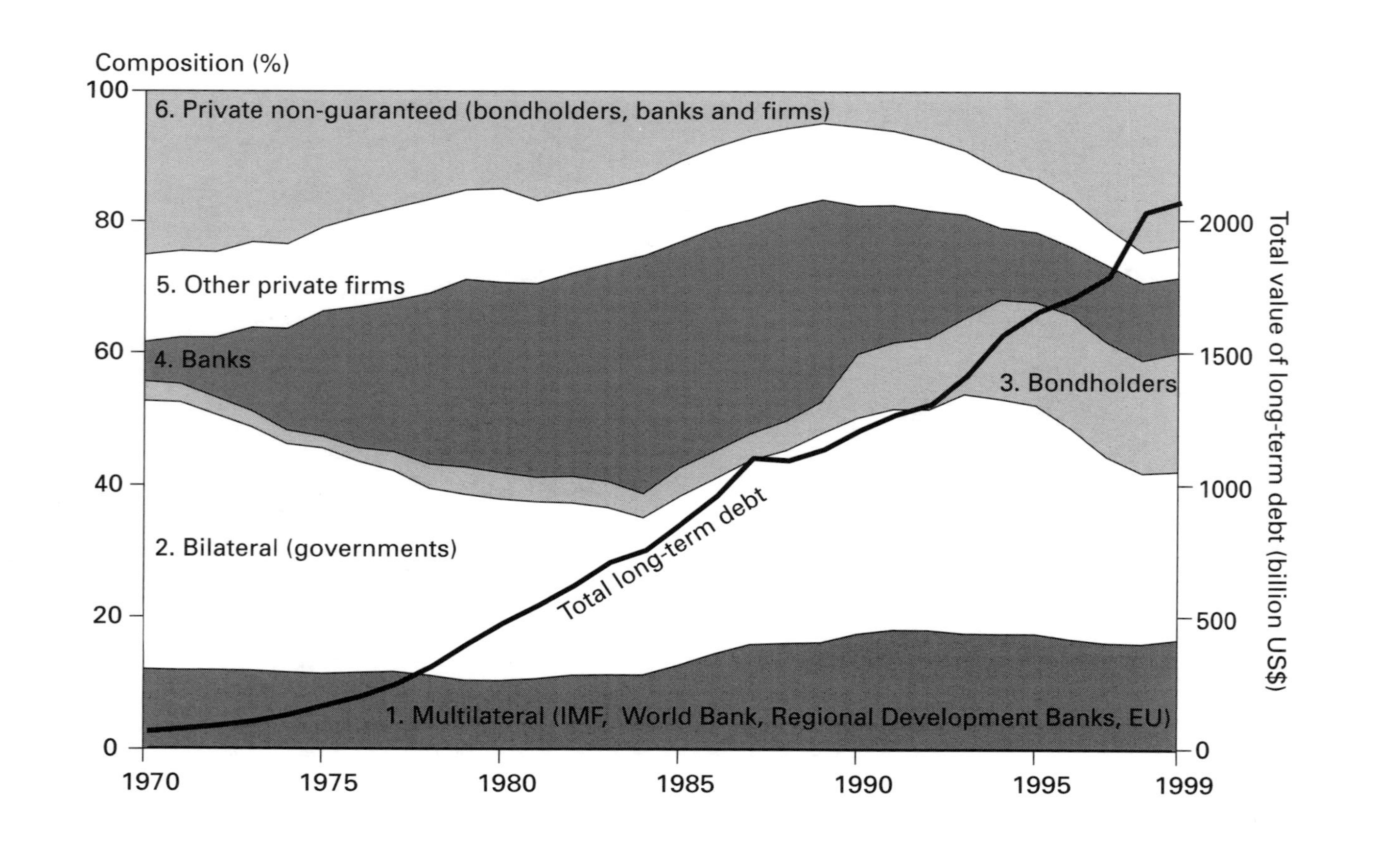
Composition (%)
100
80
60
40
20
0
6. Private non-guaranteed (bondholders, banks and firms)
5. Other private firms
4. Banks
3. Bondholders
2. Bilateral (governments)
1. Multilateral (IMF, World Bank, Regional Development Banks, EU)
Total long-term debt
Total value of long-term debt (billion US$)
2000
1500
1000
500
0
1970
1975
1980
1985
1990
1995
1999

● The rising black line (measured against the right-hand axis) shows the growth of the total size of long-term developing country debt from 1970 to 1999. The layers (measured against the left-hand axis) show the changing composition of that debt between six types of creditor.

○ This chart confirms some features of the debt already mentioned and adds some historical detail. The participation of the multilateral institutions in the debt has been relatively invariable. But this chart shows clearly the growing privatization of the debt after 1970 as banks rushed to lend their excess funds to Latin America and Asia. But to get the funds the debtor country governments had to agree to guarantee payments. The crisis of 1982, when Mexico and later most debtors sought new schedules since they could not meet existing obligations to pay, seemed for a time to threaten the stability of some of the big banks which had lent on such a large scale. The banks, however, found various ways of escaping, at relatively little cost to themselves. The slack was taken up by an expansion of bilateral lending. And the IMF began to play a much larger role as a financial supervisor of the borrowing countries in order to prevent major defaults.

During the 1990s private lenders made a major comeback. But private credits to the developing countries are different from before. Far fewer of them are bank loans; more are bonds which are not so directly threatening to the financial stability of financial institutions. And an increasing amount of the debt is not guaranteed by the governments of the debtor countries. These changes have been devised in order to produce more stability. But the experience of the late 1990s showed that they had not succeeded.

World Bank 1996; World Bank 2000d.

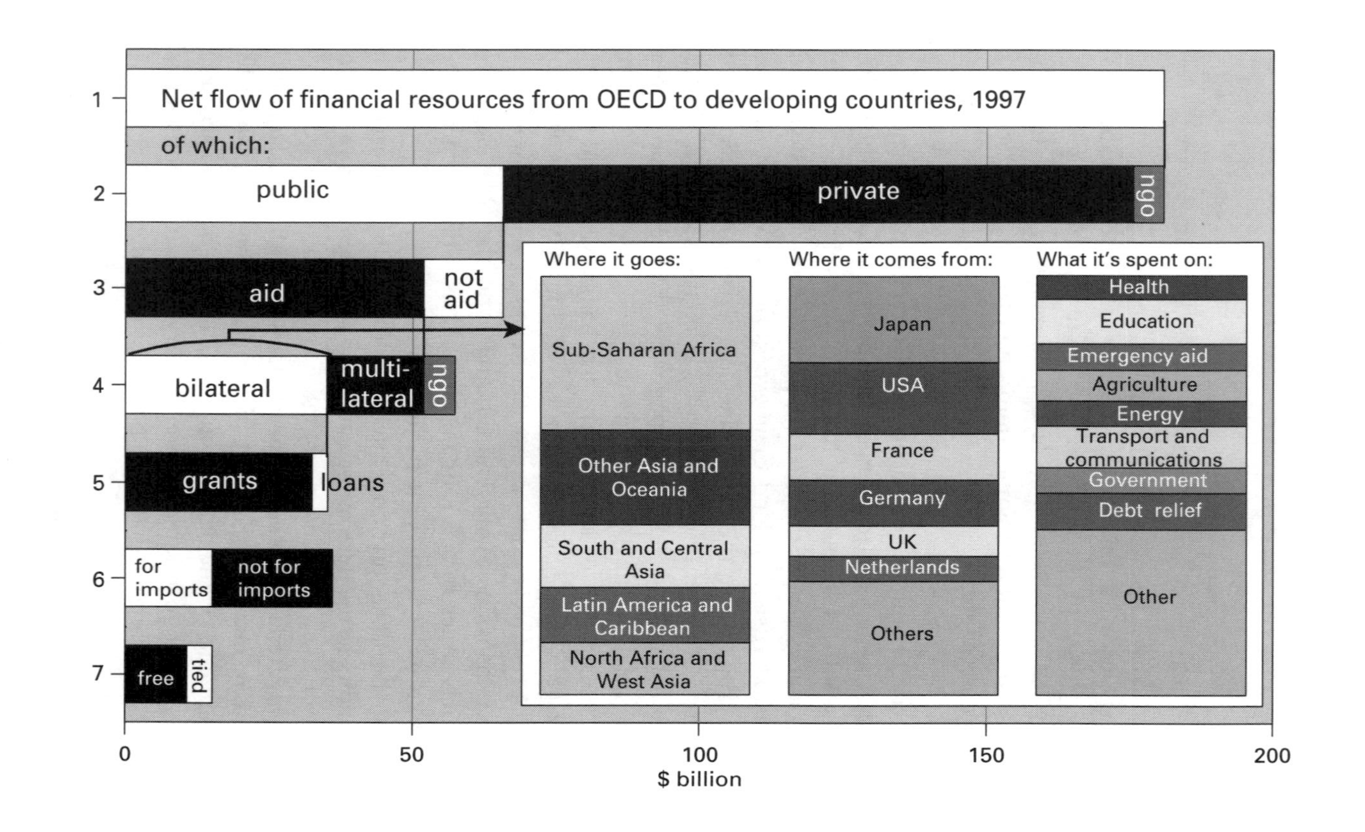
Net flow of financial resources from OECD to developing countries, 1997
of which:
public
private
ngo
aid
not aid
bilateral
multi-lateral
ngo
grants
loans
for imports
not for imports
free
tied
Where it goes:
Sub-Saharan Africa
Other Asia and Oceania
South and Central Asia
Latin America and Caribbean
North Africa and West Asia
Where it comes from:
Japan
USA
France
Germany
UK
Netherlands
Others
What it's spent on:
Health
Education
Emergency aid
Agriculture
Energy
Transport and communications
Government
Debt relief
Other
1
2
3
4
5
6
7
0
50
100
150
200
$ billion

● The graph shows the anatomy of development aid from developed to developing countries in 1997. Bar 1 shows the total flow of new resources from public, private and NGO sources. All private flows and some public flows are commercial and so do not count as development aid (bars 2 and 3). Bar 4 shows the total of aid, as usually defined, from bilateral, multilateral and NGO sources. It is divided into loans and grants (bar 5) and into the parts available and not available to spend in foreign exchange (bar 6). The available foreign exchange is divided into the part which is free to be spent anywhere and the part which is tied to purchases from the donor country (bar 7). The inset chart on the right shows the division of bilateral (state to state) aid on the basis of three different criteria: destination, source and use.

○ Private investment by capitalists from developed countries in the developed world is now noticeably more voluminous than public aid. Aid is about 60 per cent bilateral (from governments), 30 per cent multilateral (from international organizations) and 10 per cent from NGOs, although NGOs also administer a portion of public aid on behalf of governments. In principle aid is one of the main means in which world inequalities can be alleviated. But it is worth in total only about one-seventh of 1 per cent of the world's income so as a redistribution mechanism it is minuscule. Bar 6 shows that only a minority of aid actually represents money which recipient countries can spend on imports. The rest is mostly the salaries of technical assistance workers. And even a part of the foreign exchange has to be spent in the donor country and so is worth less than its face value. Aid in the form of untied foreign exchange amounts to one-fiftieth of 1 per cent of global income.

■ OECD 1999.

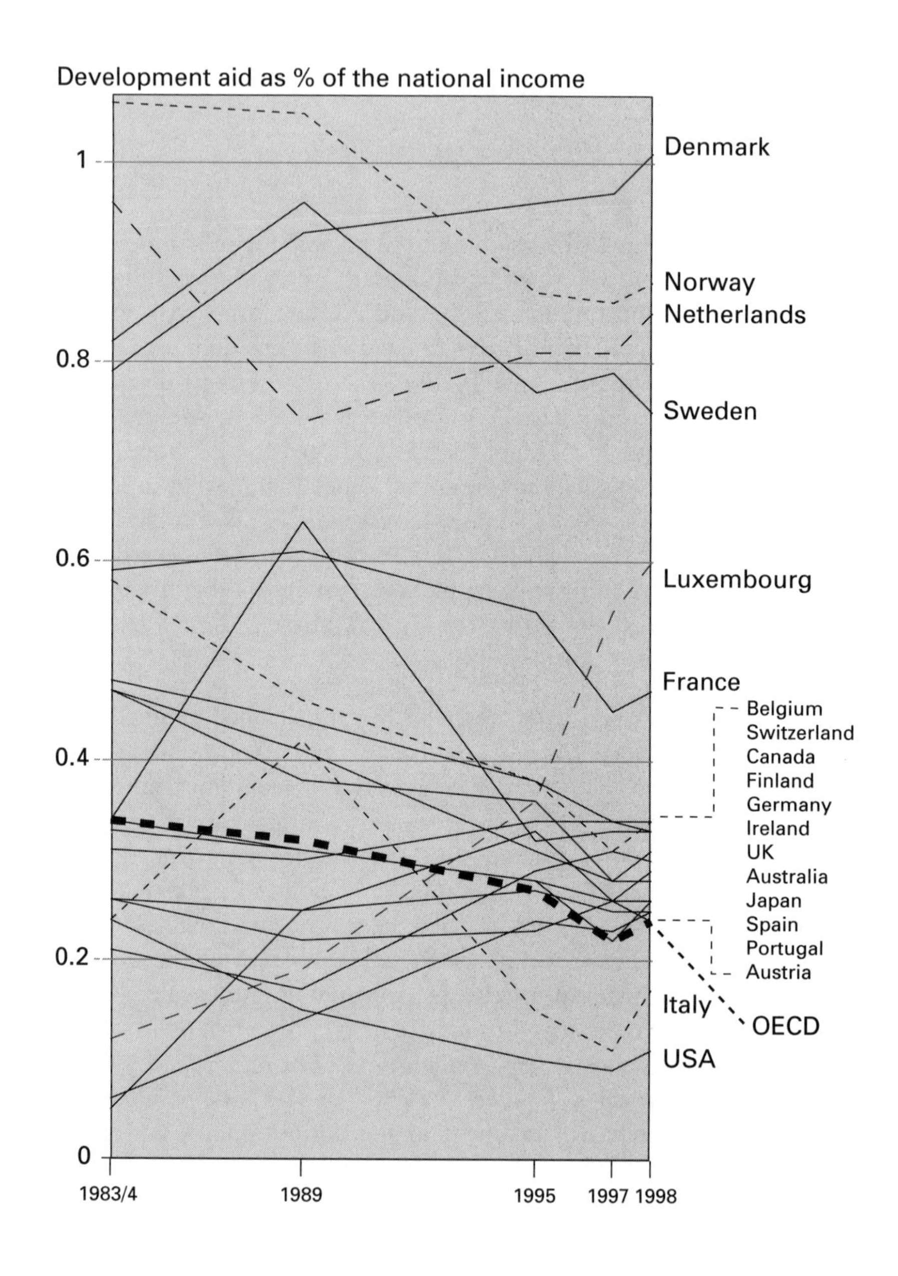

Development aid as % of the national income
1
0.8
0.6
0.4
0.2
0
1983/4
1989
1995
1997
1998
Denmark
Norway
Netherlands
Sweden
Luxembourg
France
Belgium
Switzerland
Canada
Finland
Germany
Ireland
UK
Australia
Japan
Spain
Portugal
Austria
Italy
OECD
USA

The decline of aid, 1983–1998

● This graph shows the evolution of development aid as a percentage of the national income of the donor countries in the years 1983/4 (average), 1989, 1995, 1997 and 1998. The bold dotted line is the aggregate figure for all **OECD** donor countries.

○ The best measure of a country's commitment to redress the inequalities between nations is aid as a percentage of its national income or product. The graph shows the salient features of that commitment in recent years. Four countries (Denmark, Norway, the Netherlands and Sweden) have a commitment clearly superior to the rest. They are the only countries which carry out the famous United Nations resolution which calls on developed countries to give 0.7 per cent of their national income as international aid. The overall percentage has been steadily declining over the last two decades, a victim of government austerity and anti-inflationary polices and of neo-liberal reaction against state activity in general. These figures understate the real decline since they include debt forgiveness, which is officially counted as aid and which in recent years has been one of its fastest growing components. But this is not really aid since most forgiven debt would not have been paid anyway.

The global figure is dragged down by that for the USA, the donor country which gives by far the lowest portion of its national product, where politicians of both left and right have united to oppose US aid programmes, to stop aid going to governments regarded as too progressive or too reactionary. Some convergence between other OECD countries can be observed around the figure of 0.35 per cent of national income. There is no reason to expect it to rise from that level in the near future.

■ OECD 1999.

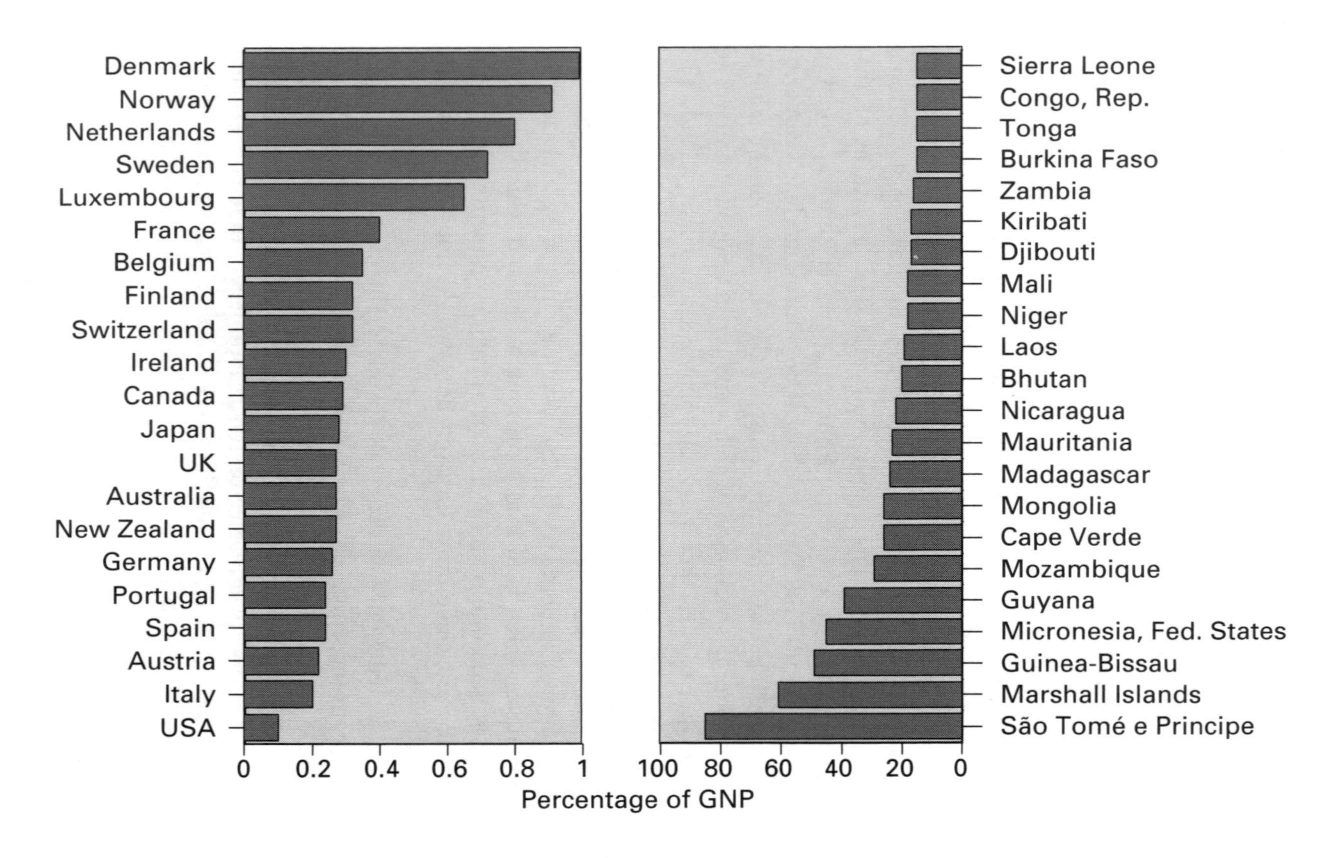
Denmark
Norway
Netherlands
Sweden
Luxembourg
France
Belgium
Finland
Switzerland
Ireland
Canada
Japan
UK
Australia
New Zealand
Germany
Portugal
Spain
Austria
Italy
USA
0
0.2
0.4
0.6
0.8
1
Sierra Leone
Congo, Rep.
Tonga
Burkina Faso
Zambia
Kiribati
Djibouti
Mali
Niger
Laos
Bhutan
Nicaragua
Mauritania
Madagascar
Mongolia
Cape Verde
Mozambique
Guyana
Micronesia, Fed. States
Guinea-Bissau
Marshall Islands
São Tomé e Principe
100
80
60
40
20
0
Percentage of GNP

Aid as a share of national income

● This and the next two similar graphs look at aid in three separate ways, from both the donors' and the recipients' points of view: in relation to national income, in absolute value and in value per head. This one shows aid by country as a percentage of each country's gross national product: for the donor countries on the left-hand side of the chart and for the recipient countries with the highest percentages on the right-hand side.

○ If aid as a share of the national income is a measure of a country's commitment to redressing world inequalities then Denmark's commitment is ten times as high as that of the USA. The typical donor supplies about 0.35 per cent of its income in the form of international aid, the result of an undecipherable mixture of generosity and self-interest.

The chart on the right shows all the recipient countries for which in 1997 aid amounted to more than one-fifth of their national income. It is therefore a measure of the relative but not the absolute size of aid. It can also be seen as a measure of dependence on aid. Aid is most important in this sense in very poor and very small countries. This is partly because much aid is now sent not for long-term development purposes but to alleviate short-term crises in the supply of food and other basic needs, the so-called humanitarian aid. And it partly reflects the fact that there is a minimum practical size to an aid package and so in very small countries this can amount to quite a large percentage of their income.

The fact that some of the poorest countries receive relatively more aid could be interpreted in two ways: as indication of a real redistribution mechanism, or as a syndrome of dependency which is hard to break because short-term solutions to poverty are sometimes themselves obstacles to long-term solutions.

■ OECD 1999.

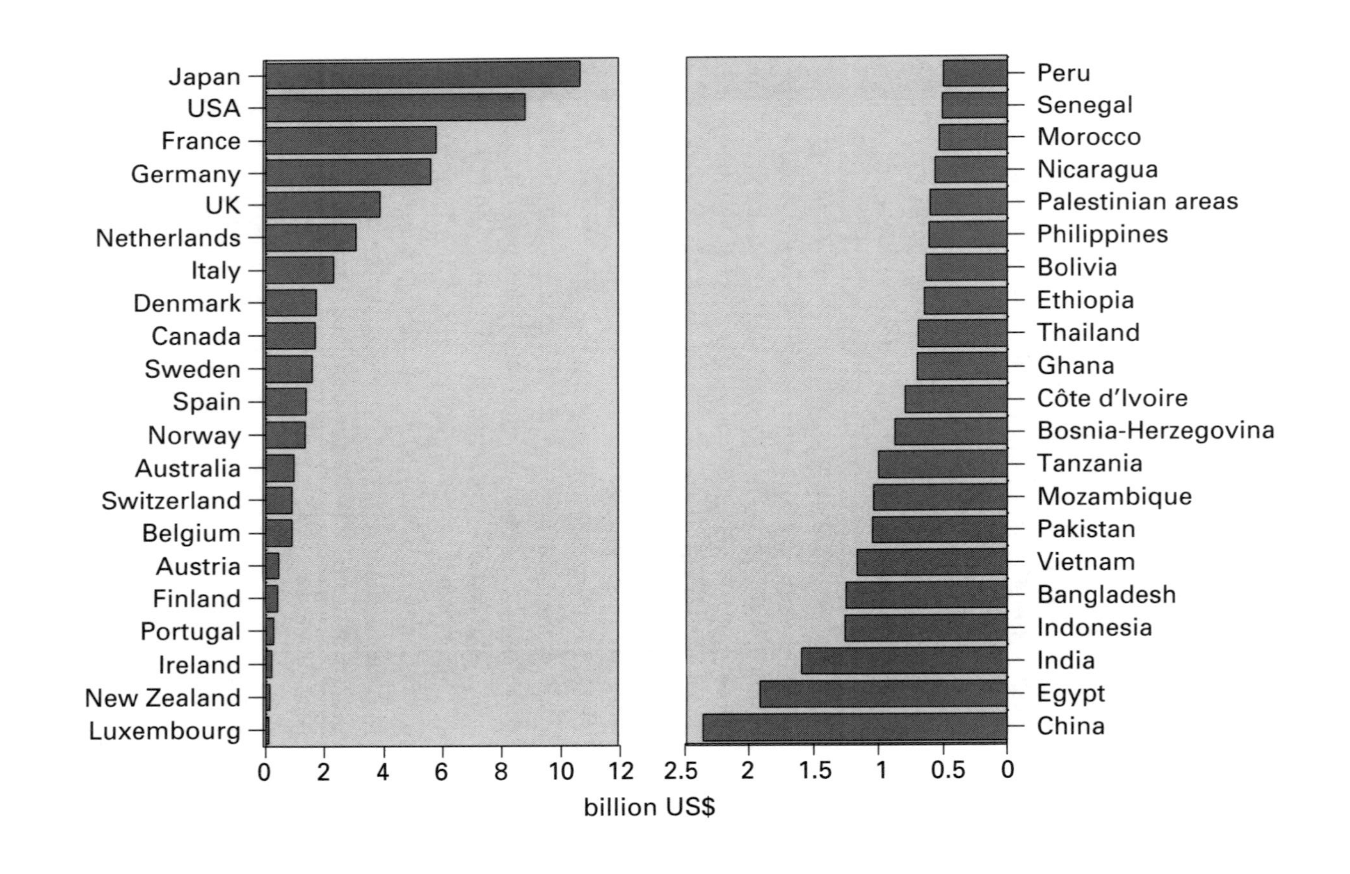
Japan
USA
France
Germany
UK
Netherlands
Italy
Denmark
Canada
Sweden
Spain
Norway
Australia
Switzerland
Belgium
Austria
Finland
Portugal
Ireland
New Zealand
Luxembourg
0
2
4
6
8
10
12
2.5
2
1.5
1
0.5
0
billion US$
Peru
Senegal
Morocco
Nicaragua
Palestinian areas
Philippines
Bolivia
Ethiopia
Thailand
Ghana
Côte d'Ivoire
Bosnia-Herzegovina
Tanzania
Mozambique
Pakistan
Vietnam
Bangladesh
Indonesia
India
Egypt
China

● This graph shows the absolute amounts of aid distributed in 1997. On the left appear all the donor countries and on the right those recipient countries which received more than $500 million.

○ Japan has become the largest distributor of aid. This reflects both the declining enthusiasm of US political leaders for aid and the rising desire of Japan to play a more active world role. To some extent the facts of the right hand of the chart are reflected in the left-hand part. Logically the big recipients are the countries most favoured by the big donors. Japan's aid efforts have been mostly directed towards countries in Asia, and in addition a few in Latin America. The USA has traditionally given a high volume of aid to Egypt (in an effort to bolster order in the Middle East) and has in recent years given aid as part of the US-sponsored peace processes; that accounts for the presence of the embryonic Palestinian state and of Bosnia and Herzegovina on this list. One or two of the poorest countries are on the list because they have managed to attract aid from quite a wide range of sources; these include, Nicaragua, Mozambique, Tanzania and Bangladesh.

China has recently become one of the largest recipients of aid from developed countries. In this case the predominant motive is certainly one which dominates many aid flows: to prepare the way for more intense economic relations between the donor and the recipient country, which in this case is a rapidly growing country with the second largest national product in the world. Aid is thus seen by many governments as a catalyst for profitable commercial relations.

■ OECD 1999.

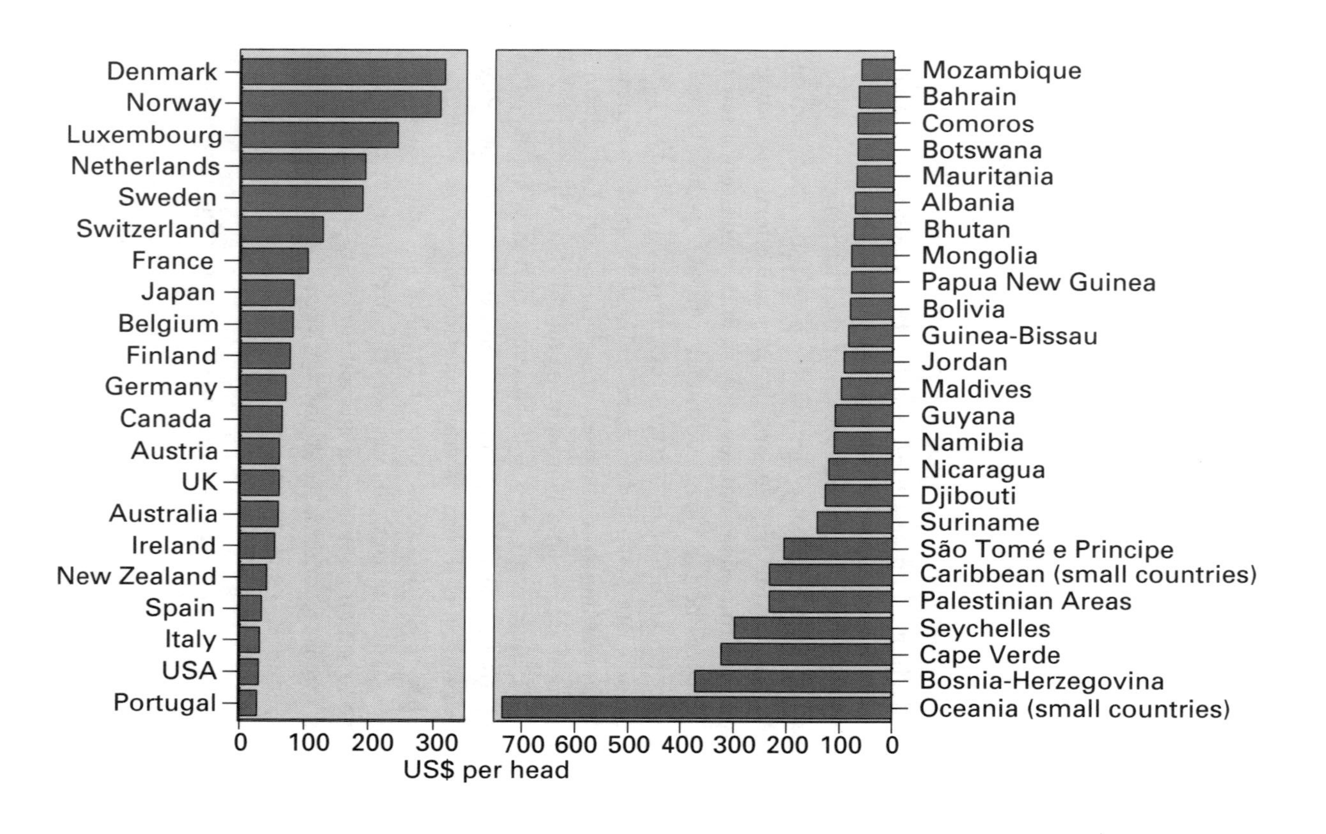
Denmark
Norway
Luxembourg
Netherlands
Sweden
Switzerland
France
Japan
Belgium
Finland
Germany
Canada
Austria
UK
Australia
Ireland
New Zealand
Spain
Italy
USA
Portugal
0
100
200
300
US$ per head
700
600
500
400
300
200
100
0
Mozambique
Bahrain
Comoros
Botswana
Mauritania
Albania
Bhutan
Mongolia
Papua New Guinea
Bolivia
Guinea-Bissau
Jordan
Maldives
Guyana
Namibia
Nicaragua
Djibouti
Suriname
São Tomé e Principe
Caribbean (small countries)
Palestinian Areas
Seychelles
Cape Verde
Bosnia-Herzegovina
Oceania (small countries)

● The last of these three graphs shows aid in terms of dollars per head for both donors and all the recipients which receive more than $50 per person.

○ The order of donor countries by dollars of aid per head is very similar to the order according to percentage of the national income. The same four countries (plus Luxembourg) are the most generous. The least generous in this sense are the USA and some southern European countries.

This measure shows that in general countries with very small populations receive most aid per head. The reason is that aid packages almost inevitably have a minimum size and this can be very large in relation to the population of a small island. Very few of the recipients which appear in this list also appear in the previous graph: in other words the countries which receive the largest amount of total aid tend not to receive very large amounts of aid per head. In view of this it is worth drawing attention to the four countries which are on both lists: Nicaragua, the Palestinian areas, Bosnia and Herzegovina, and Bolivia. Two of these appear because aid is part of an internationally sponsored peace process; the other two are the poorest countries in South and Central America, respectively.

There is no close relationship between the amount of aid which recipients receive and the amount of debt service which they pay. But, as a final observation about aid, it should be noticed that its total amount is very much less than the amount of debt service paid by the developing countries (› 107). For some countries, therefore, aid in effect does not even pay for the service of the debt.

■ OECD 1999.

The environment VII

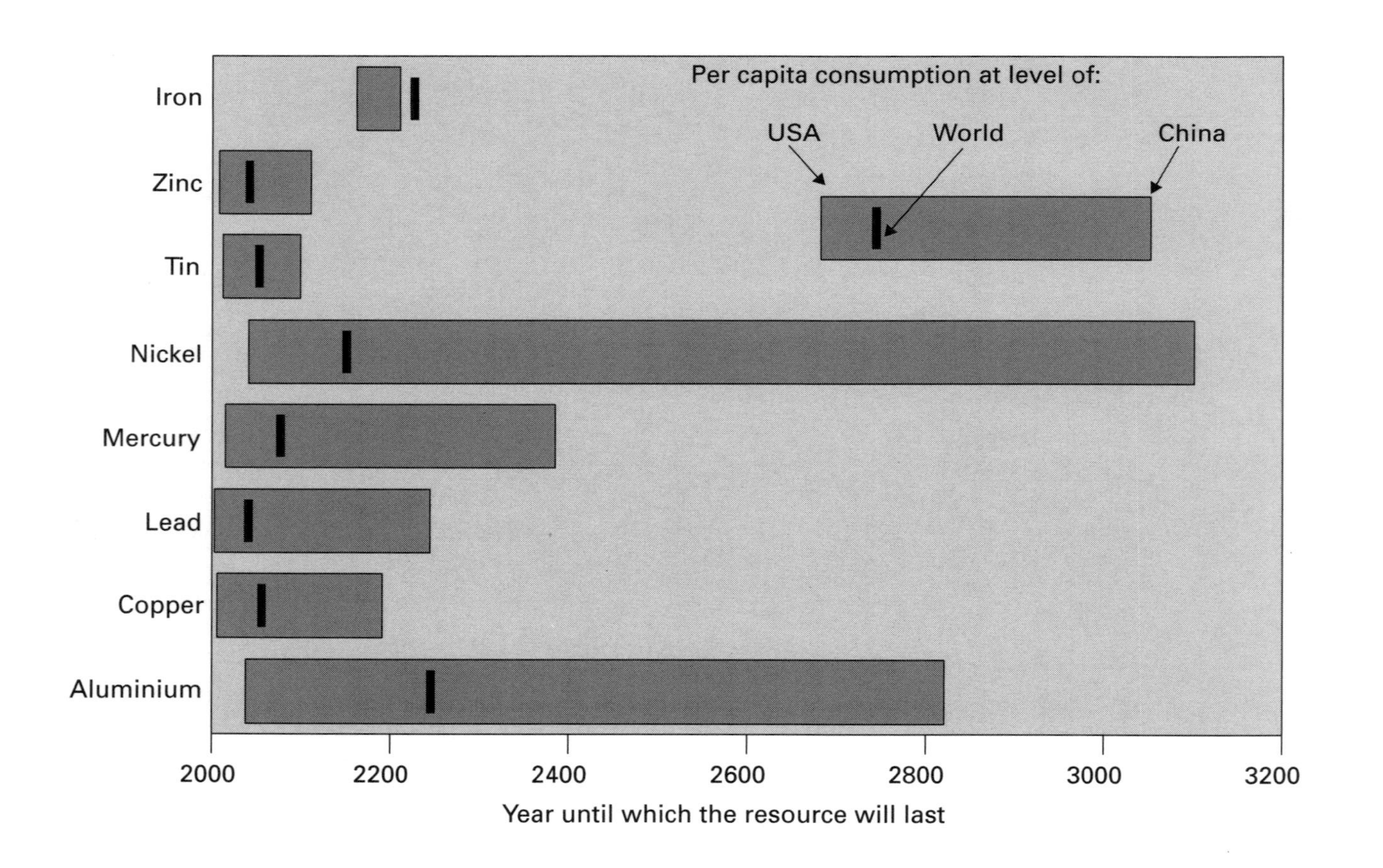

Per capita consumption at level of:
USA
World
China
Iron
Zinc
Tin
Nickel
Mercury
Lead
Copper
Aluminium
2000
2200
2400
2600
2800
3000
3200
Year until which the resource will last

● The graph measures the amount of time which known reserves of the mentioned natural resources will last if the whole world consumed them at the present rate per head of the USA (the left edge of the bar) and China (the right edge of the bar). The vertical line shows the time reserves will last at present average world consumption levels.

○ Inequality has great relevance to the question of whether human life and economic activity are sustainable. The ecological perspective argues that the developed way of life is not sustainable because it overuses non-renewable resources and creates multiple forms of contamination. If less developed countries become more like developed ones, then the limits of sustainability will be reached sooner. The world as a whole is incapable of sustaining a situation in which all people live on average like its richest occupants today. According to this argument the burden which humans impose upon the earth is unequal: the lives of the inhabitants of developed countries are more damaging because they use more materials and they create more contamination. This graph shows in a very simplified form the first part of that argument. Given the known reserves of the eight non-renewable materials shown here, the year to which they would last varies according to their level of use per head. On the present average level of consumption per head of the whole world, they would last until the year indicated by the vertical line. The bars show that US levels of consumption, if generalized, would exhaust some important materials very soon but that they would last longer at Chinese levels of consumption. If in this and other respects the US model of development is unsustainable for the world, then other less resource-hungry models will have to be found.

■ WRI 1998.

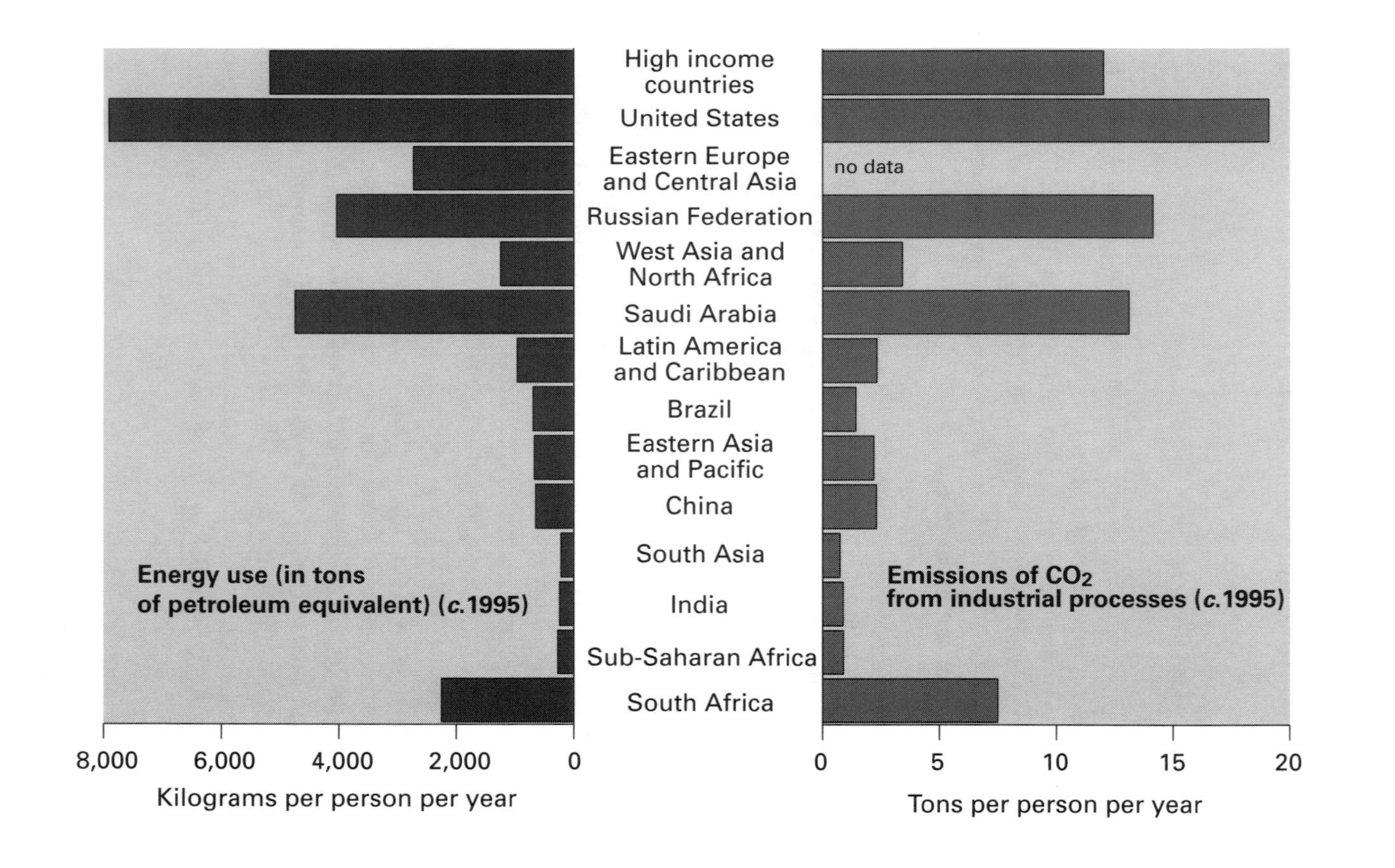

High income countries
United States
Eastern Europe and Central Asia
Russian Federation
West Asia and North Africa
Saudi Arabia
Latin America and Caribbean
Brazil
Eastern Asia and Pacific
China
South Asia
India
Sub-Saharan Africa
South Africa
Energy use (in tons of petroleum equivalent) (c.1995)
8,000
6,000
4,000
2,000
0
Kilograms per person per year
no data
Emissions of CO_2 from industrial processes (c.1995)
0
5
10
15
20
Tons per person per year

● The two charts show the levels per head of energy use (left) and carbon dioxide emissions (right) for some representative countries and country groupings.

○ The second ecological charge against modern economic development is that it leads to excessive emissions of dangerous pollutants. Many of these have noxious local effects but others spread their effect more widely. Concern centres on the emission of carbon dioxide, which is believed by most climatologists to lead to rising CO_2 concentrations in the earth's atmosphere; this in turn causes global warming, whose consequences may include a rising sea level, greater climatic extremes and thereby serious negative effects on the possibilities of producing food and other aspects of human life. In other words this is part of an argument that the patterns of development of the last two hundred years are not sustainable. Modern economic growth is in some respects self-negating in that it destroys the basis for its own continuation. This argument supplements the many aspects of human life mentioned earlier in this book which lead to a critique of conventional development from the point of view of its desirability. The ecological critique questions its continued possibility. And the inequality between nations and among people is at the heart of both these debates.

These graphs show the extremely unequal levels of contribution to the problem of excess carbon dioxide emission, usually linked especially to the growth in the use of fossil-based energy. The relative levels of energy use, therefore, are closely related to international differences in CO_2 emissions; and they are both, of course, very closely related to differences in levels of income (‹ 1, › 99).

■ WRI 1998.

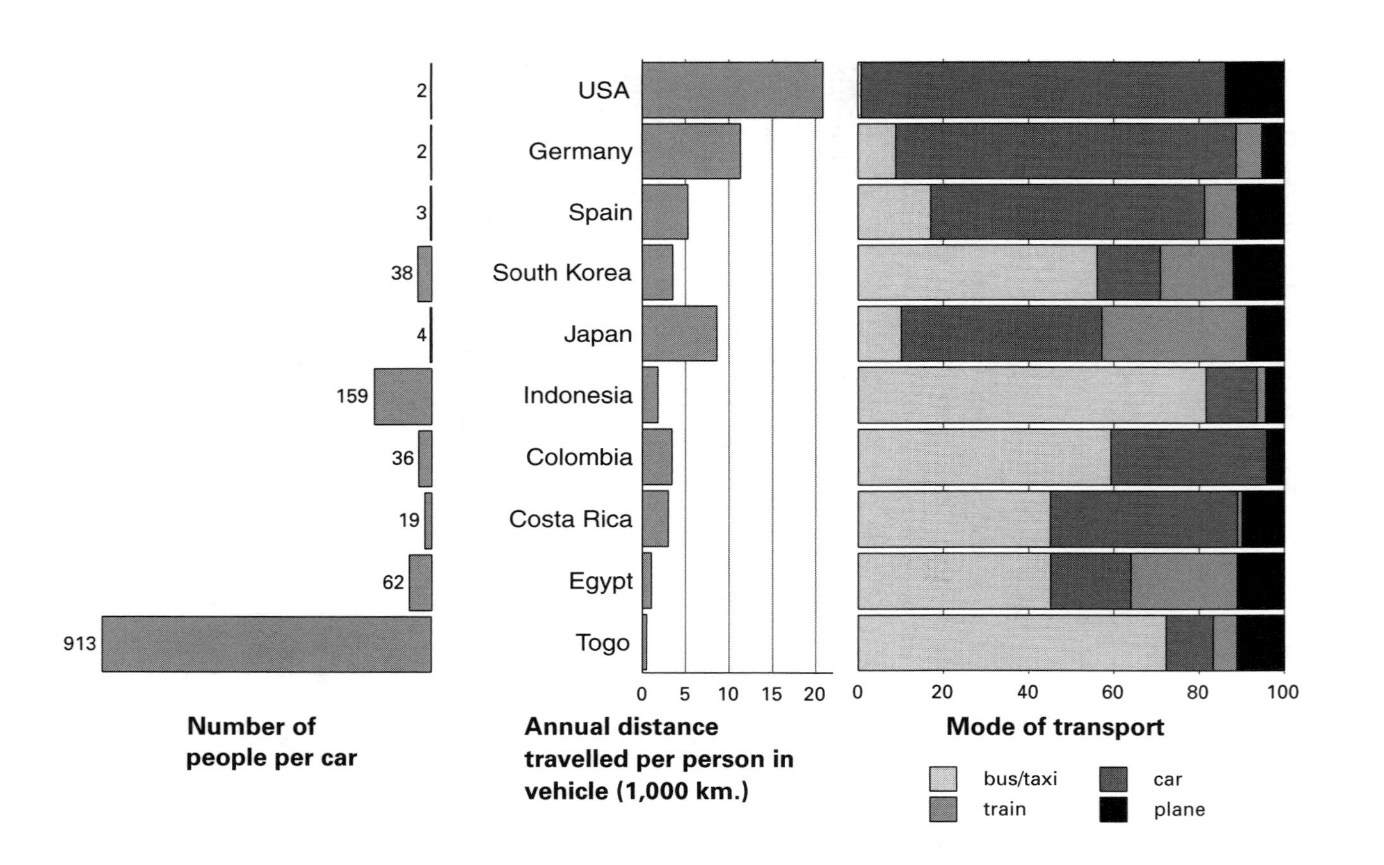
2
2
3
38
4
159
36
19
62
913
Number of people per car
USA
Germany
Spain
South Korea
Japan
Indonesia
Colombia
Costa Rica
Egypt
Togo
0 5 10 15 20
Annual distance travelled per person in vehicle (1,000 km.)
0 20 40 60 80 100
Mode of transport
bus/taxi
car
train
plane

Contrast in transportation

● The three charts show aspects of international differences in transport: the relative number of automobiles (left), the amount which people on average travel (centre), and the mode of transport which they use (right). The data are for the early 1990s.

○ One of the main causes of rising energy use and rising CO_2 emissions is transport. These ten countries, representative of countries at different levels of income, show various aspects of this relationship. People in richer countries tend to travel much more than people in poor countries (as a result of living further from their work and places of entertainment and of taking more holidays away from home). There also seems to be a systematic difference in the relative use of modes of transport between countries at different economic levels. What is not observable here is that people in poorer countries travel a great deal more by foot and by bicycle than those in richer countries. What the right-hand chart does show is that for motorized transport people are much more inclined to travel in mass public forms of transport (buses, trains and taxis) in poorer countries and much more inclined to travel by car (which uses more energy and creates more contamination per person) in the richer countries. The relative importance of air travel is similar between some rich and poor countries since it reflects not only income differences but also differences in the size and accessibility of countries. As often in these questions, the USA is the extreme case: its citizens travel on average over 20,000 kilometres a year, over 80 per cent of them in cars. If the ratio of cars to people that exists in the USA existed in the whole world, then there would be not the 600 million cars which exist in the world today but 3,100 million.

■ WRI 1996; WRI 1998.

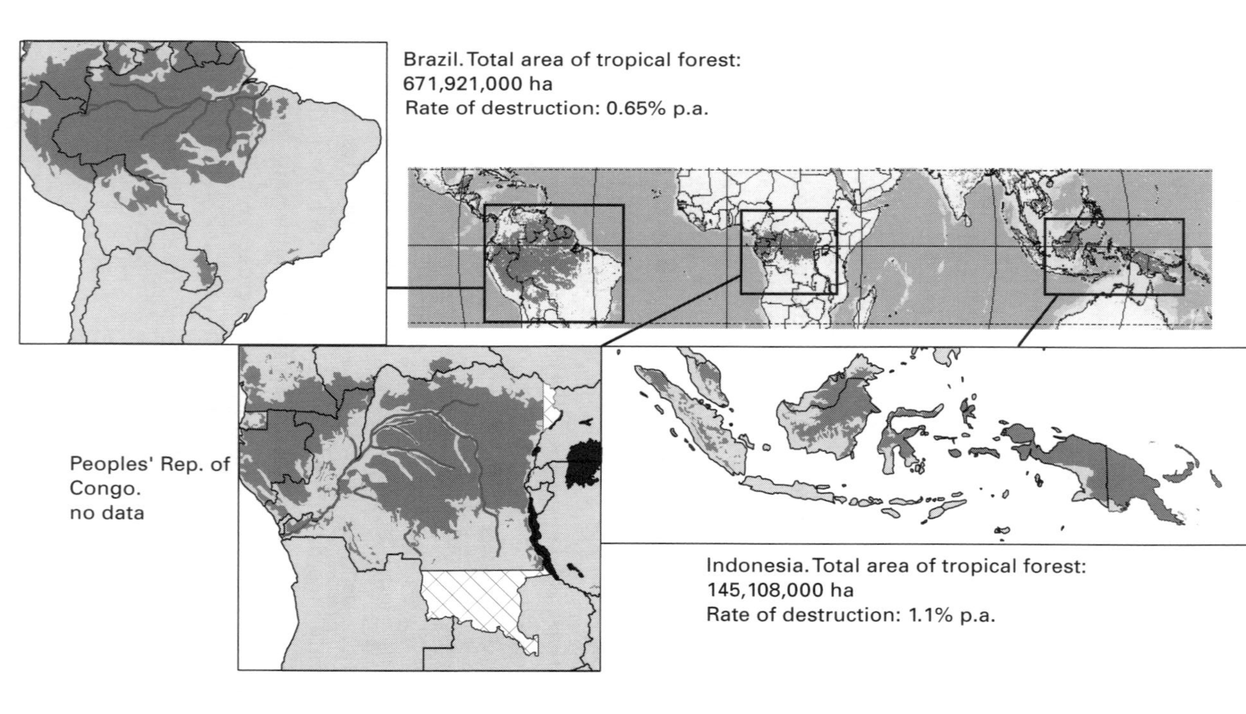

Brazil. Total area of tropical forest:
671,921,000 ha
Rate of destruction: 0.65% p.a.
Peoples' Rep. of
Congo.
no data
Indonesia. Total area of tropical forest:
145,108,000 ha
Rate of destruction: 1.1% p.a.
main areas of tropical rain forest

area without data

● On the map of the tropics and the three more detailed maps remaining areas of tropical rainforest are marked in grey.

○ The quantity of carbon dioxide which can be naturally absorbed depends very crucially on vegetation. Forests, and particularly tropical rainforests, play a central role. The problem of global warming, therefore, is seen as a combination of those factors which increase emissions with those factors which reduce absorption.

In the second category is the reduction of forest area in general but in particular the reduction of the tropical rainforest, regarded as the lungs of the planet. These maps show how concentrated is this resource. Large areas of tropical forest exist in only a very few countries: Brazil and a few neighbouring countries in South America, the Democratic Republic of the Congo, and two or three other central-west African states, and in Indonesia.

World inequalities impinge in various ways in this equation. Riches produce the market for expensive tropical hardwoods, which is partly responsible for giving a monetary incentive for excessive felling of rainforests. And poor countries with tropical woods often have an incentive to exploit them excessively in order to attract investment by foreign companies and expand foreign exchange receipts. Also poverty and landlessness in the tropical countries lead to the destruction of forests to provide cultivable land even in countries where plenty of unused, unforested land exists but is unavailable to poor farmers.

■ World Conservation Monitoring Centre 1999.

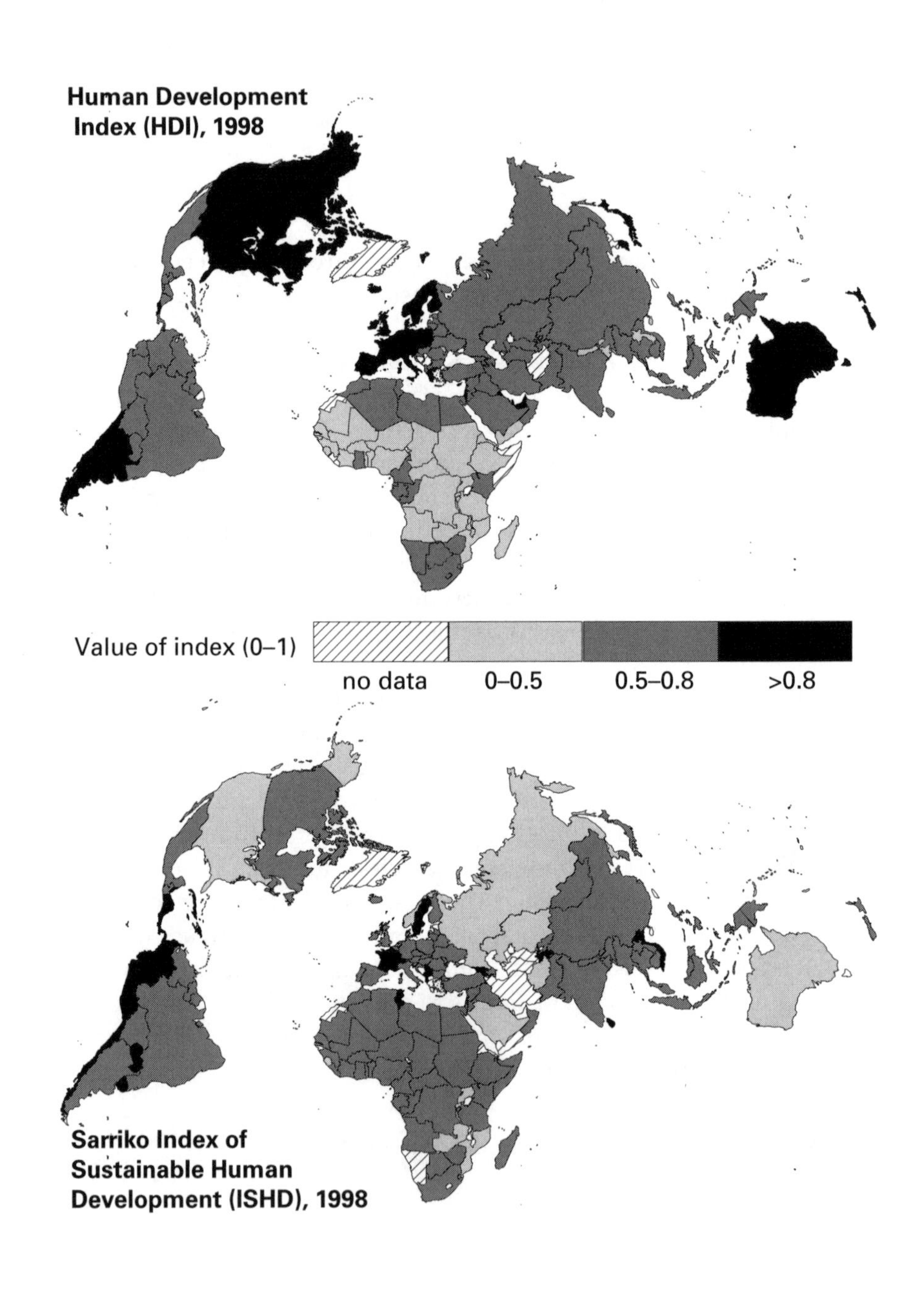
Human Development Index (HDI), 1998
Value of index (0–1)
no data
0–0.5
0.5–0.8
>0.8
Sarriko Index of Sustainable Human Development (ISHD), 1998

A measure of sustainable human development

● The upper map is a repeat of the view of the world according to the HDI (graph 21). The lower one is based on a simple index of sustainable human development (named after the author's workplace). It has only two variables – life expectancy at birth, a positive variable to represent human development, and emissions per head of carbon dioxide, a negative variable to represent environmental contamination. The formula for each country follows the methodology of the HDI. Like the HDI, the SISHD can vary between 0 and 1.

○ International organizations have gone beyond measuring development and economic welfare through differences of income alone. They have produced indices of the quality of life and human capacity like the HDI. But so far they have not produced indices which take the environment and sustainability into account. The result shown here may explain why.

Measures which take into account the quality of life produce somewhat different results from those based on income levels alone. But the order remains similar and exceptions are few. But when a negative variable is added representing the amount of environmental contamination, as it is in this simple home-made index, everything is transformed. The most advanced countries are those which were in intermediate positions according to previous measures. The least advanced are either those which have always been considered least developed (because they have very low life expectancy) or those which were considered most developed (since they have very high levels of carbon dioxide emissions). The most advanced states are Costa Rica, Sri Lanka and Albania. And the very least advanced include Sierra Leone and the United States.

■ UNDP 1999.

Refugees and migration VIII

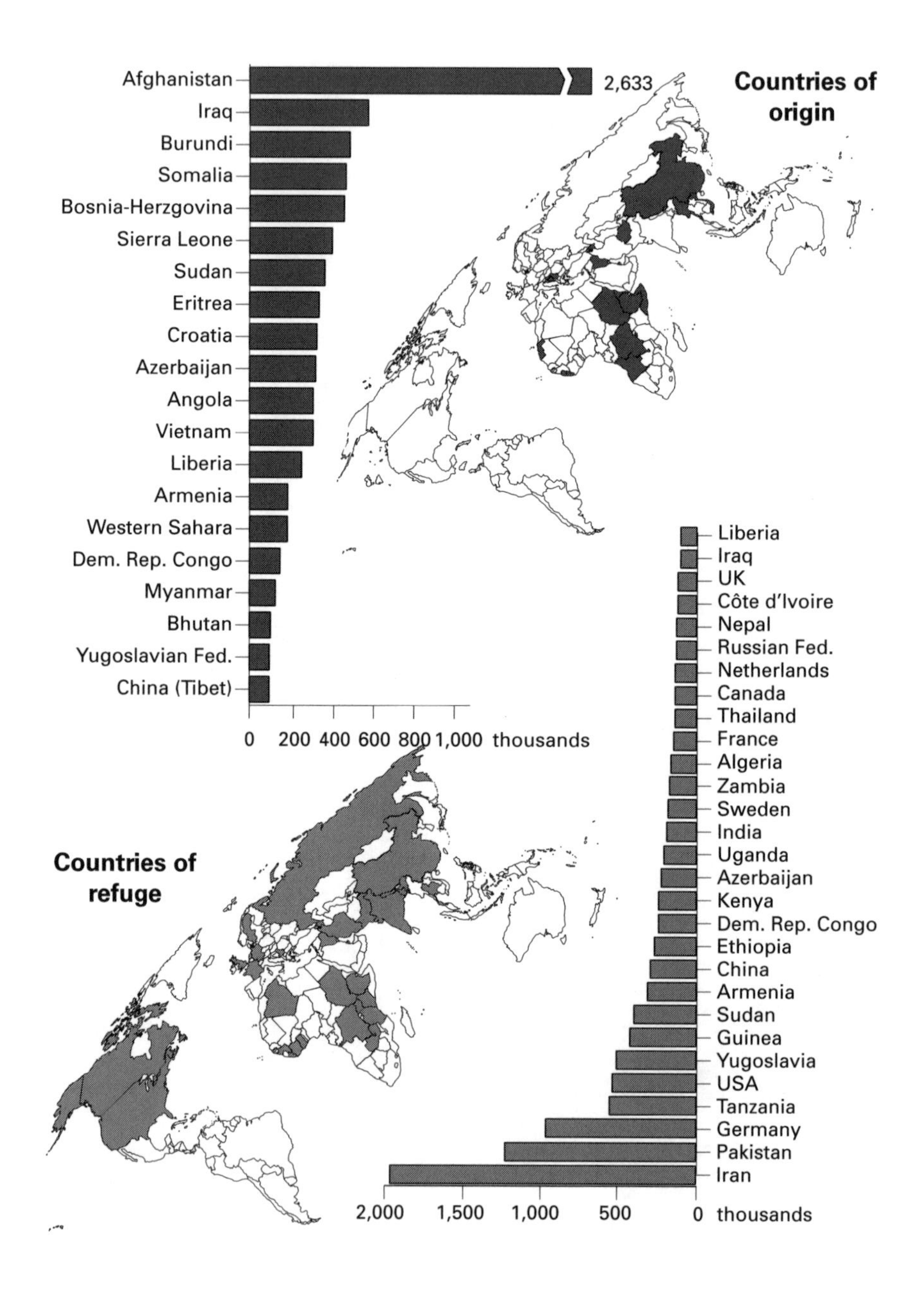
Countries of origin
Afghanistan
2,633
Iraq
Burundi
Somalia
Bosnia-Herzgovina
Sierra Leone
Sudan
Eritrea
Croatia
Azerbaijan
Angola
Vietnam
Liberia
Armenia
Western Sahara
Dem. Rep. Congo
Myanmar
Bhutan
Yugoslavian Fed.
China (Tibet)
0 200 400 600 800 1,000 thousands
Countries of refuge
Liberia
Iraq
UK
Côte d'Ivoire
Nepal
Russian Fed.
Netherlands
Canada
Thailand
France
Algeria
Zambia
Sweden
India
Uganda
Azerbaijan
Kenya
Dem. Rep. Congo
Ethiopia
China
Armenia
Sudan
Guinea
Yugoslavia
USA
Tanzania
Germany
Pakistan
Iran
2,000 1,500 1,000 500 0 thousands

The origin and destination of refugees, 1998

100

● The upper and lower bar charts show the main countries of origin and residence of forced international migrants in 1998. Countries with more than 100,000 forced emigrants or immigrants are shown in the charts and marked on the map.

○ It was estimated that at the end of 1998 around 13 million people in the world were forced international migrants – asylum seekers and refugees from hunger, war and oppression, all resident beyond the borders of the country of which they are citizens. And about 5 million more have been forced to migrate internally.

The most salient fact about the pattern of forced migration is that most forced emigrants come from poor countries and most of them find their refuge in other poor, usually neighbouring, countries. So they tend to leave without any resources and to live in extremely poor refuges. Forced migrations therefore often intensify poverty and deprivation. Even when migrants receive their elementary needs through international aid, they usually cannot establish themselves in their country of refuge. Their situation is often dominated by the idea, in practice often a myth, that they will return to a normalized homeland. But millions of people have lived in refugee camps for decades, proving the falsity of this assumption.

People forced to migrate come from a relatively small number of countries, though the number of countries of refuge is rather more extensive, as can be seen in these graphs. In 1998 Afghanistan was still the single most important country of origin of refugees and, reflecting this, the two most important countries of refuge were Iran and Pakistan, Afghanistan's neighbours. If the European Union sheltered as many refugees in relation to its population as Iran they would number 11 million!

■ UNHCR 1999.

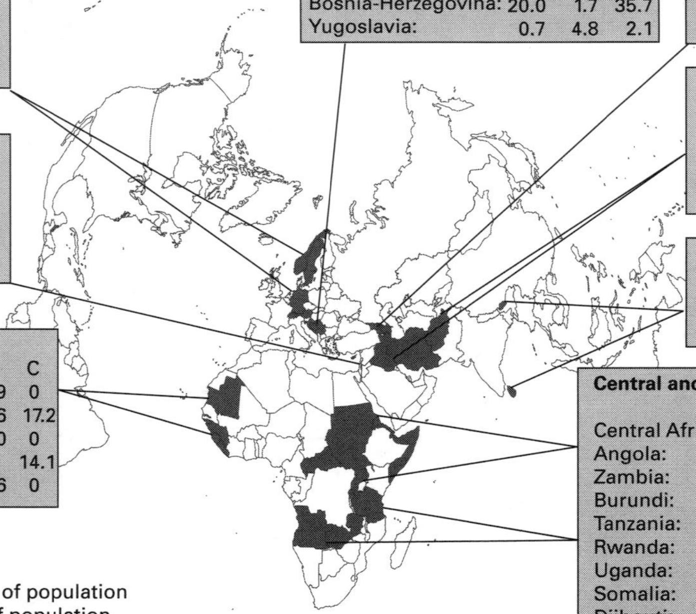

European refuges

	A	B	C
Norway:	0	1.0	0
Sweden:	0	2.0	0
Germany:	0	1.6	0
Switzerland:	0	1.8	0
Austria:	0	1.1	0

The Palestinian refugees

	A	B	C
Palestine:	50	--	--
Jordan:	0	28.2	0
Lebanon:	0	8.2	0
Syria:	0	2.2	0

West African civil wars

	A	B	C
Mauritania:	2.8	0.9	0
Guinea-Bissau:	0.7	0.6	17.2
Guinea:	0	6.0	0
Sierra Leone:	8.6	0	14.1
Liberia:	8.9	3.6	0

The Balkan wars

	A	B	C
Croatia:	7.0	0.6	1.5
Bosnia-Herzegovina:	20.0	1.7	35.7
Yugoslavia:	0.7	4.8	2.1

Wars in the Caucasus

	A	B	C
Georgia:	2.7	0	5.1
Azerbaijan:	3.9	2.3	7.3
Armenia:	5.0	8.2	0

West Asian wars and revolutions

	A	B	C
Iraq:	2.7	0.5	?
Iran:	0.2	3.2	0
Afghanistan:	10.5	0	1.3

Other crises

	A	B	C
Sri Lanka:	0.5	0	3.3
Bhutan:	14.3	0	0

Central and East African crises

	A	B	C
Central African Republic:	0	1.4	0
Angola:	2.7	0.1	0
Zambia:	0	1.8	0
Burundi:	7.7	0.4	0
Tanzania:	0	1.8	0
Rwanda:	0.9	0.4	7.9
Uganda:	0	1.0	0
Somalia:	5.4	0	0
Djibouti:	0.5	3.8	0
Eritrea:	9.1	0.1	0
Sudan:	1.4	1.4	0

A Outward refugees as % of population
B Inward refugees as % of population
C Internally displaced as % of population
(where one or more of these is greater than 1%)

The main locations of forced migration, 1998

● This map shows all those countries where any one or more of three statistics exceeds 1 per cent of the population. The three are the number of refugees from that country, the number of refugees from elsewhere in that country and the number of internally displaced people in that country, all as a percentage of the population. The figures are shown in the boxes.

○ A few countries appear on this map because they, more or less reluctantly, offer refuge to relatively large numbers of people from elsewhere (Norway, Sweden, Germany, Switzerland and Austria in Europe; Guinea, Central African Republic, Zambia, Tanzania and Uganda in Africa). Others appear because internal social, political and ethnic conflicts have generated a large number of refugees (Somalia and Bhutan). Others appear because their conflicts have caused a large amount of forced migration within their own borders (Guinea-Bissau, Rwanda, Sri Lanka). Several of the countries highlighted here appear for more than one of the above reasons. They are the source of forced migration and also the recipients of forced migration from elsewhere. This occurs either because there have been serious internal conflicts in two neighbouring countries between which refugees mutually flow (for instance Liberia and Sierra Leone or Iran and Iraq) or because they are part of a more regional conflict in which states do not coincide with ethnic groups, as in Palestine or in the manifold conflict where multiple refugee populations have been created.

If the European Union had the highest figures shown here for these three measures, 160 million of its citizens would be refugees elsewhere, it would have 100 million refugees from elsewhere, and 120 million of its citizens would be internally displaced.

UNHCR 1999; Palestinian Refugee Research Net 2000.

The demographic composition of refugees in UNHCR camps

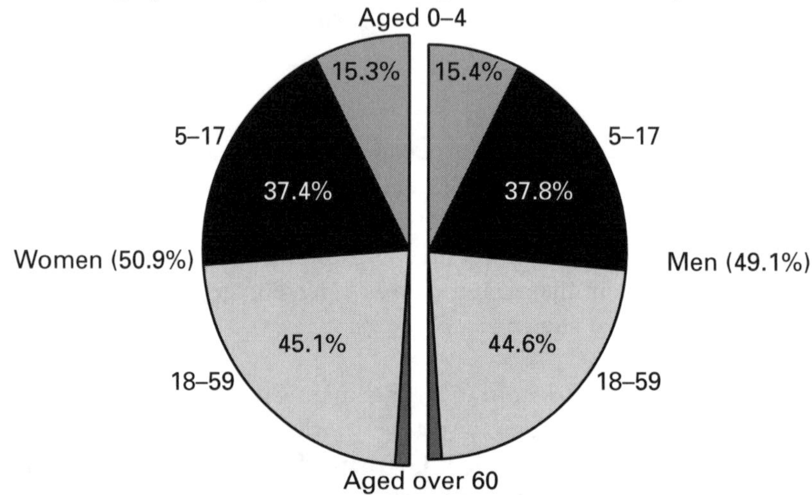

The composition of refugee populations by country of origin

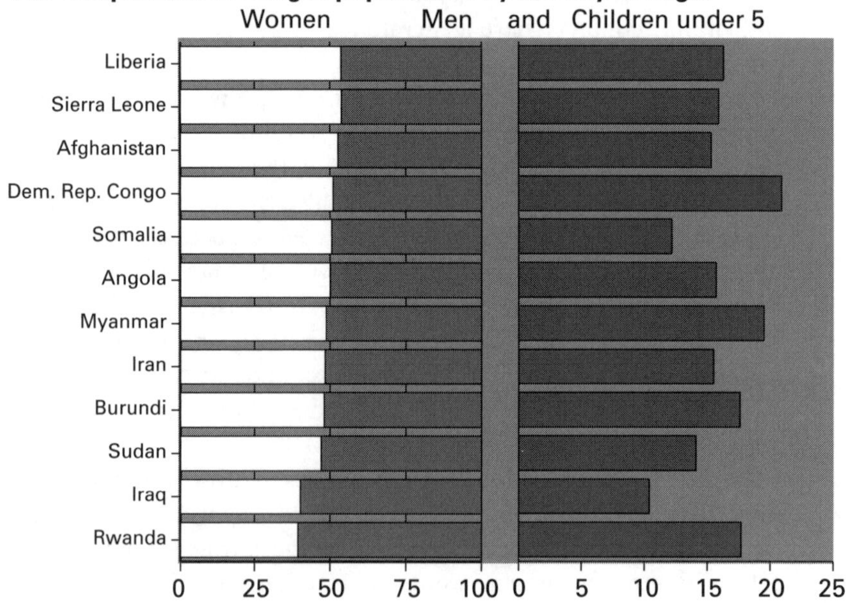

● These graphs give demographic information about those refugees who in 1998 lived in refugee camps run by the United Nations High Commission for Refugees. The upper pie charts show the age and sex distribution of the refugees in these camps and the lower bar charts show the shares in the refugee populations in certain countries of men, women and children under 5 years old.

○ This demographic analysis covers a little over 4 million refugees out of the estimated world total of 13 million. It suggests that there are important demographic differences between the refugee population and the world population in general. In the first place, a majority, albeit a small one, of refugees are women while the majority of the world population are men (‹26). Second, a very tiny proportion of refugees are over 60 years old, considerably less than in the world population as a whole. And third, the majority of refugees are under 17 years old, a considerably higher percentage than in the world population as a whole.

The special demographic characteristics of refugees are also partly characteristics of the countries from which they come. Since these are nearly all poor countries the small percentage of old people and a large percentage of children among refugees is not surprising. But refugee populations are not simply a demographic microcosm of the countries from which they come. In a number of countries special features of the conflict have effects on the demography of the refugees. Liberia, Afghanistan, Somalia and Angola all have a majority of men in their populations, but the majority of refugees from those countries are women. In Rwanda and Burundi the opposite is the case.

■ UNHCR 1999.

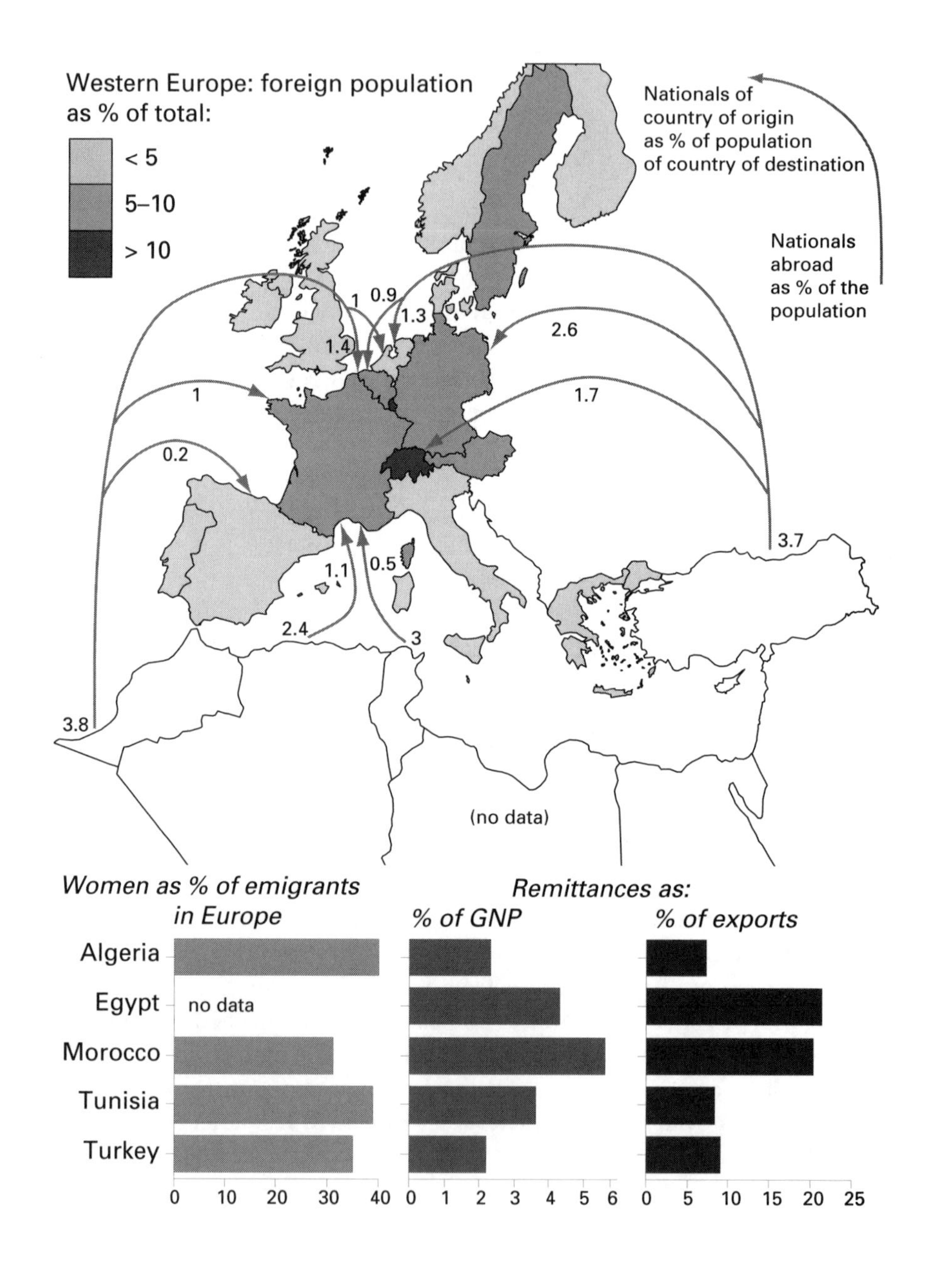

Western Europe: foreign population as % of total:
< 5
5–10
> 10
Nationals of country of origin as % of population of country of destination
Nationals abroad as % of the population
1
0.9
1.3
1.4
2.6
1
1.7
0.2
3.7
1.1
0.5
2.4
3
3.8
(no data)
Women as % of emigrants in Europe
Remittances as:
% of GNP
% of exports
Algeria
Egypt
no data
Morocco
Tunisia
Turkey
0
10
20
30
40
0
1
2
3
4
5
6
0
5
10
15
20
25

● This is the first of five graphs which provide some basic information about the role of migration in the world today. The map shows the proportion of non-nationals (not necessarily immigrants) in the populations of Western European countries and data about the migration to those countries from neighbouring countries around the Mediterranean.

○ Emigration and inequality are connected in many ways. Emigration is partly a reaction to inequality. It can alleviate the inequality if the aims of those who emigrate are achieved, or worsen it if they are not. And it expresses and reveals major differences in the way different groups of humans live, both materially and culturally, but also in terms of the civil and political rights which they enjoy. Reactionary political forces in the developed countries use and foment prejudices against immigrants as a centrepiece of their political strategy.

In Europe the proportion of foreign citizens in resident populations ranges from between 1 per cent of the population to over 15 per cent, with the average around 8 per cent. The nationalities with greatest presence are Turkish, Algerian, Moroccan and Tunisian (except in the UK, where the most numerous foreign nationalities are Indian, Pakistani and Bangladeshi). These figures do not include naturalized immigrants. A significant majority of Turkish and Magrebi migrants in Europe are men, which suggests that some of them regard their migration as temporary. But the proportion of women is growing – implying that migrant populations are becoming more settled.

The bar charts show that the remittances of migrants is extremely important for the economies of the sending countries.

■ SOPEMI 1999; World Bank 1999a.

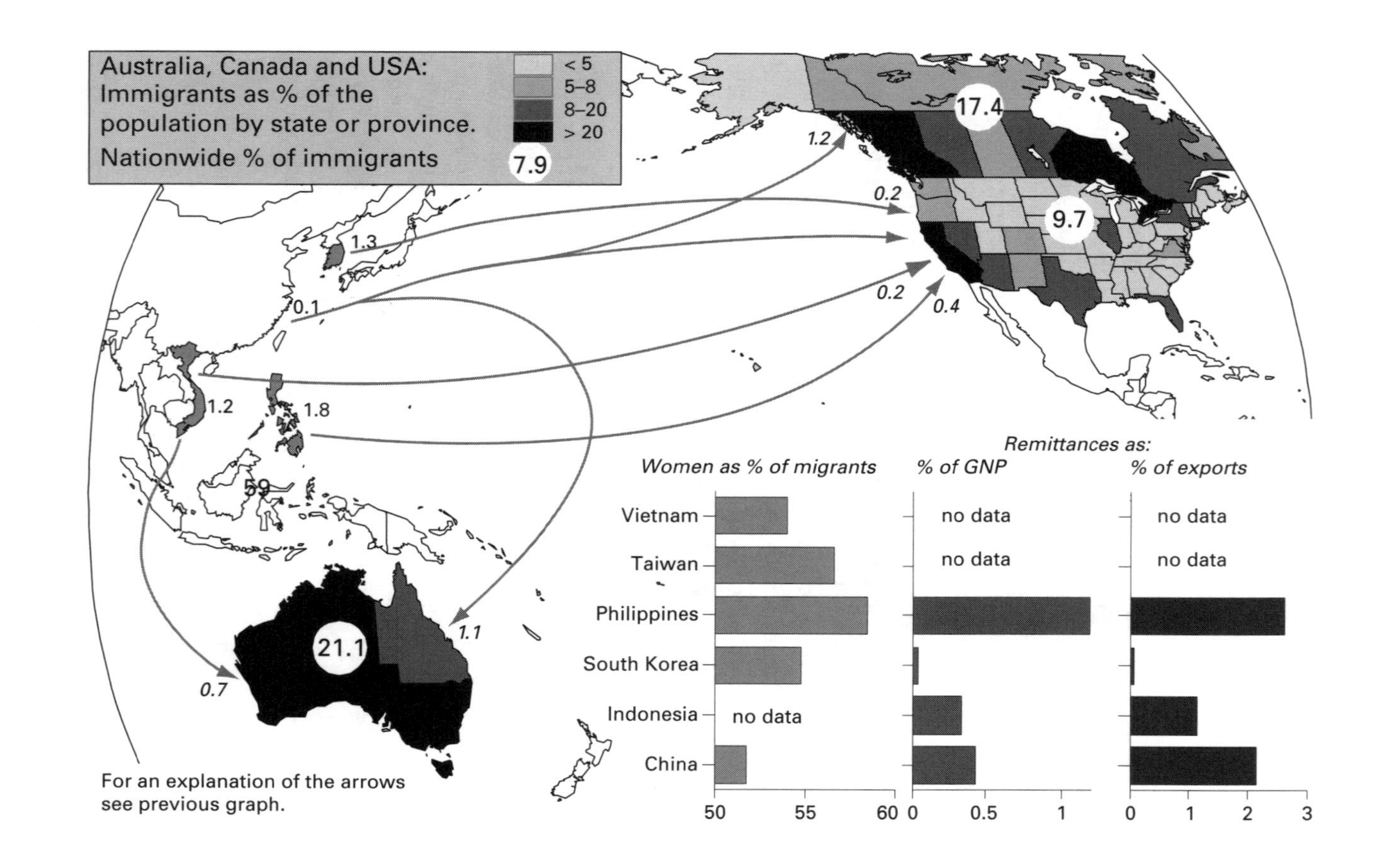

For an explanation of the arrows see previous graph.

● The map shows the percentage of immigrants (those born outside the country) in the USA, Canada and Australia both nationally (black figure in white circle) and by state or province. The arrows show the direction of flows of migration from four Asian countries with information about their magnitude. The bar charts add data about the sexual composition of migrant populations and about remittances.

○ Changes in immigration law in the USA, Canada and Australia in recent decades have expanded the possibility of immigration to those countries from Asia. More settlers in Australia now come from Asia than from English-speaking countries, and in the USA and Canada Asian immigrants form a substantial proportion of the total. A clear majority of Asian immigrants into the USA are women, suggesting that in general it is permanent family migration, though the figure for the Philippines includes large numbers of women from that country who migrate to work in domestic service. The number of people of Filipino origin in the USA is equal to nearly 2 per cent of the population of the country of origin. For South Korea the figure is about 1.2 per cent. South Korea is an instance of several countries in the world with significant outward and inward migration flows. Chinese migrants from the People's Republic as well as from Taiwan and Hong Kong have established sizeable communities in North America, especially in California and British Columbia.

Immigrants are a considerably larger share of the population in Australia and Canada than in the USA. There is also more regional similarity in the number of immigrants in Australia and Canada than in the USA, where high concentrations of immigrants are found especially in coastal and frontier states.

■ SOPEMI 1999; AUB 1991.

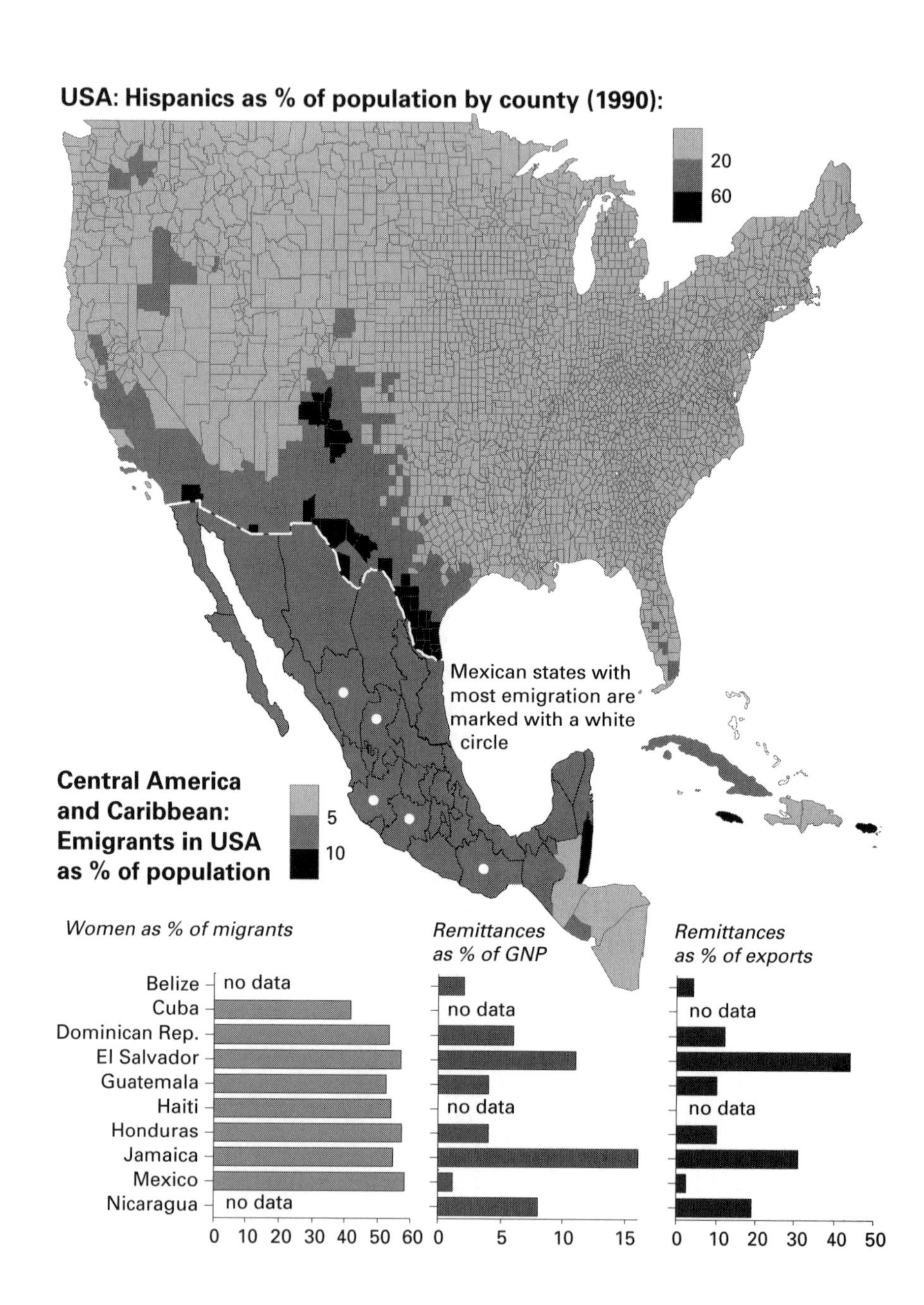
USA: Hispanics as % of population by county (1990):
20
60
Mexican states with most emigration are marked with a white circle
Central America and Caribbean: Emigrants in USA as % of population
5
10
Women as % of migrants
Remittances as % of GNP
Remittances as % of exports
Belize
Cuba
Dominican Rep.
El Salvador
Guatemala
Haiti
Honduras
Jamaica
Mexico
Nicaragua
no data
0 10 20 30 40 50 60
0 5 10 15
0 10 20 30 40 50

● The map shows the density of 'Hispanics' in the population of the USA by county, the states of Mexico from which most immigrants come, and data about the relative size of migration from different Central American countries. The bar charts show the percentage of women and the importance of migrants' remittances.

○ Nearly one half of legal immigrants (55–60 per cent of them women) and the enormous majority of illegal ones (mostly men) come from the Spanish-speaking countries of Central and South America, above all from Mexico. The 'Hispanic' population of the USA (almost 12 per cent of the total) is the result of more than a century of movement across the US–Mexican frontier, which has accelerated in the last three decades. US residents born in Mexico amount to 7 per cent of the population of their country of origin. The map shows the porousness of this, the clearest and in some ways the most important of the borders between the rich and poor parts of the world.

The ability of migrants to work in the USA and to send back money to their families has become an indispensable aspect of economic survival in the countries of origin. Remittances are equal to as much as 44 per cent of exports in El Salvador, and 31 per cent in Jamaica. Without them such economies would be devastated. There is continuous pressure from some political forces in the USA for more rigid controls and, ironically, since the fall of the Berlin wall, more physical fences have been raised along the US–Mexican border. But they have little effect on the flow of migrants. Most recent US measures to curtail immigration have perverse effects, increasing pre-emptive migration or applications for citizenship by foreign residents.

■ SOPEMI 1999.

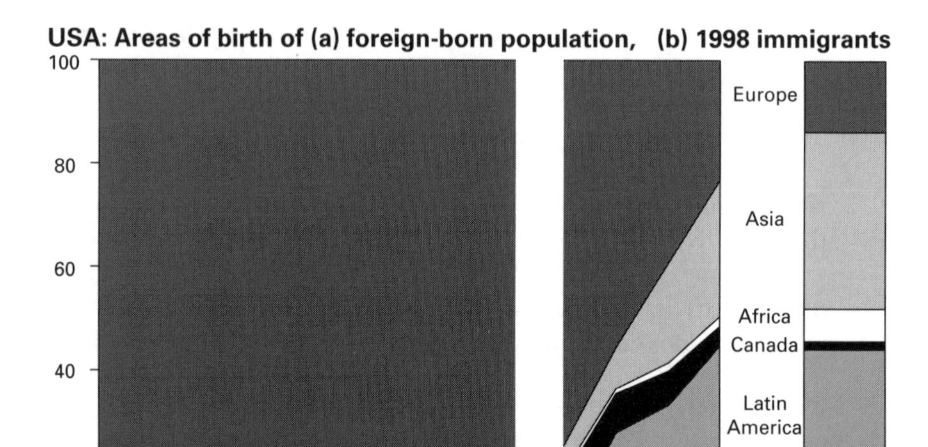
USA: Areas of birth of (a) foreign-born population, (b) 1998 immigrants
100
80
60
40
20
0
1850 1860 1870 1880 1890 1900 1910 1920 1930 1960 1970 1980 1990
1998
Europe
Asia
Africa
Canada
Latin America
(Mexico)

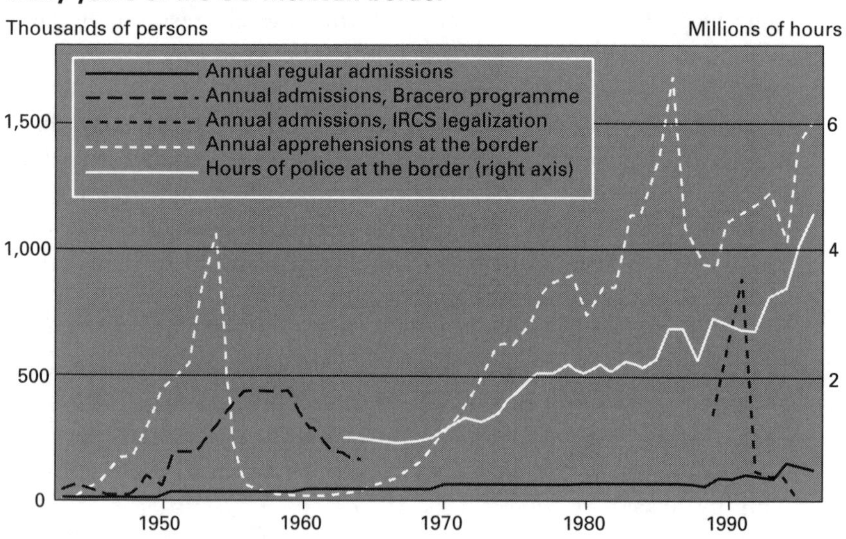
Sixty years of the US-Mexican border
Thousands of persons
Millions of hours
Annual regular admissions
Annual admissions, Bracero programme
Annual admissions, IRCS legalization
Annual apprehensions at the border
Hours of police at the border (right axis)
1,500
1,000
500
0
6
4
2
1950
1960
1970
1980
1990

● The upper chart shows the continental origin of the foreign-born population of the USA by decade from 1850 to 1990 (the break between 1930 and 1960 is due to the absence of comparable figures), and the areas of origin of legal immigrants in 1998. The lower chart shows information about legal and illegal migration across the US–Mexican border.

○ Before 1965 US immigration law was based on the notion of the national quota, calculated according to the percentage national origin of the existing US population, not counting African-Americans. This racist act was repealed as part of the civil rights wave of the 1960s and replaced with an overall quota without restrictions on region. The result, as the upper chart clearly shows, has been a complete transformation of the nature of immigration to the USA. Once predominantly from the developed countries of Europe, it now comes almost entirely from the developing countries of Asia, Africa and Latin America.

Immigration from Mexico is the largest and the most conflictive component of US immigration. Resources devoted to policing the border have fluctuated greatly over the last fifty years (see lower chart). The general rise in apprehensions of illegal border crossers may reflect an increase in illegal migration, though it also simply mirrors the number of border patrols. The intake of legal migrants is not high but periodic amnesties (such as the 1986 Immigration Control and Reform Act) encourage illegal migrants to believe that they will one day become legal residents. A border which is wide open for the passage of goods and money is hard to close definitively to the passage of people.

■ US Bureau of the Census 2000; Inter-American Dialog 2000.

Migrant workers' remittances, 1997

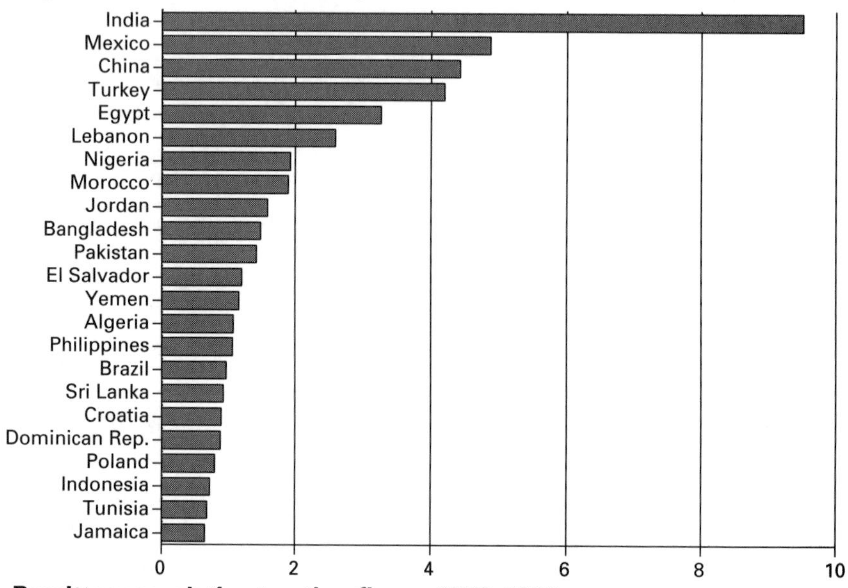

Remittances relative to other flows, 1970–1998

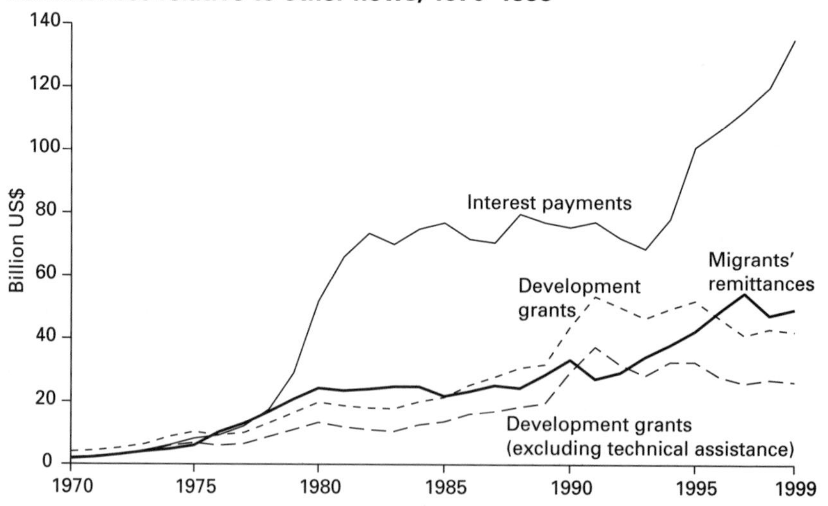

Migrants' remittances: more important than aid

107

● The upper chart shows the total amount of registered migrants' remittances to the indicated countries in 1997. The bottom chart compares declared remittances with other financial flows between 1984 and 1998.

○ In general it can be said that there are three ways in which the income gaps between rich and poor countries could be closed: convergence in national development levels, economic aid and migrants' remittances. The data summarized here show that remittances already constitute an important redistribution mechanism. These individual payments by citizens of developing countries amount to a much greater transfer of resources to their countries than everything done by rich country states, NGOs and international organizations, although this is hardly ever acknowledged in international debate. Moreover, remittances are greatly underestimated in the official figures (since some of them are clandestine and some in kind) while aid figures tend to be inflated due to tying and the inclusion of such items as the forgiveness of unpayable debt (‹ 90–94). It is sometimes argued that remittances do not contribute much to development since they go directly to emigrants' families, who spend them on consumption. In fact they are almost certainly better spent than aid, much of which feeds corruption and bureaucracy. There is evidence that a high proportion of remittances are spend on education.

Remittances provide a part of the foreign exchange which pays interest on the external debt, that is why a number of governments of developing countries have encouraged temporary migration and take measures to oblige migrants to repatriate as much of their foreign earnings as possible.

■ World Bank 1999a; World Bank 2000d.

Repression and discrimination IX

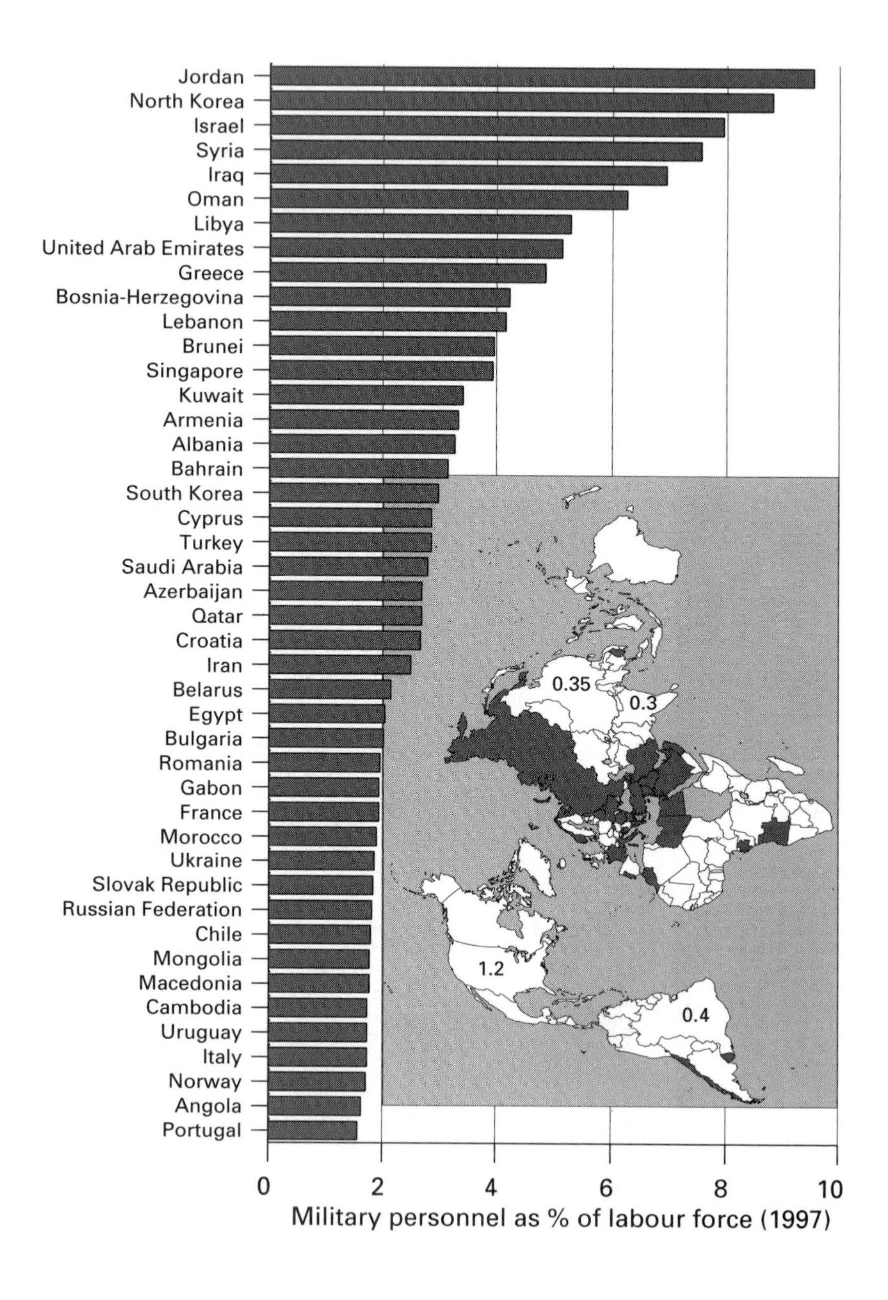

Jordan
North Korea
Israel
Syria
Iraq
Oman
Libya
United Arab Emirates
Greece
Bosnia-Herzegovina
Lebanon
Brunei
Singapore
Kuwait
Armenia
Albania
Bahrain
South Korea
Cyprus
Turkey
Saudi Arabia
Azerbaijan
Qatar
Croatia
Iran
Belarus
Egypt
Bulgaria
Romania
Gabon
France
Morocco
Ukraine
Slovak Republic
Russian Federation
Chile
Mongolia
Macedonia
Cambodia
Uruguay
Italy
Norway
Angola
Portugal
0.35
0.3
1.2
0.4
0
2
4
6
8
10
Military personnel as % of labour force (1997)

● The bar chart shows all those countries where members of the armed forces and official paramilitary personnel amount to more than 1.5 in every hundred members of the labour force. These countries are shaded on the map to show their geographical concentration. The equivalent figure for four major countries, the USA, Brazil, India and China, are indicated on the map for comparison.

○ This list of countries makes sobering if not very surprising reading. It shows how the political tensions of a few parts of the world are translated into the structure of human activity. In first place is the area conventionally, if Eurocentrically, known as the Middle East – the eastern Mediterranean and the Persian/Arabian Gulf. Virtually every country in this region appears in this list of the most militarized countries. So also do both North and South Korea, the countries of the Balkans, most countries of the former Soviet Union, Angola and Chile.

The relationship between armies and inequality is a complex one but it certainly exists at many levels. Armies play a major role in policing international inequalities. They may also be important in policing internal inequalities, especially where the military has direct political power. Armies are themselves creators of inequality since they nearly always have privileged access to material resources of all kinds. When food is scarce, for instance, it is usually available to soldiers. This is necessary for governments in order to maintain their political power but it is also an incentive to recruitment: families with members in the military may be able to survive more easily than other families.

■ World Bank 1999a.

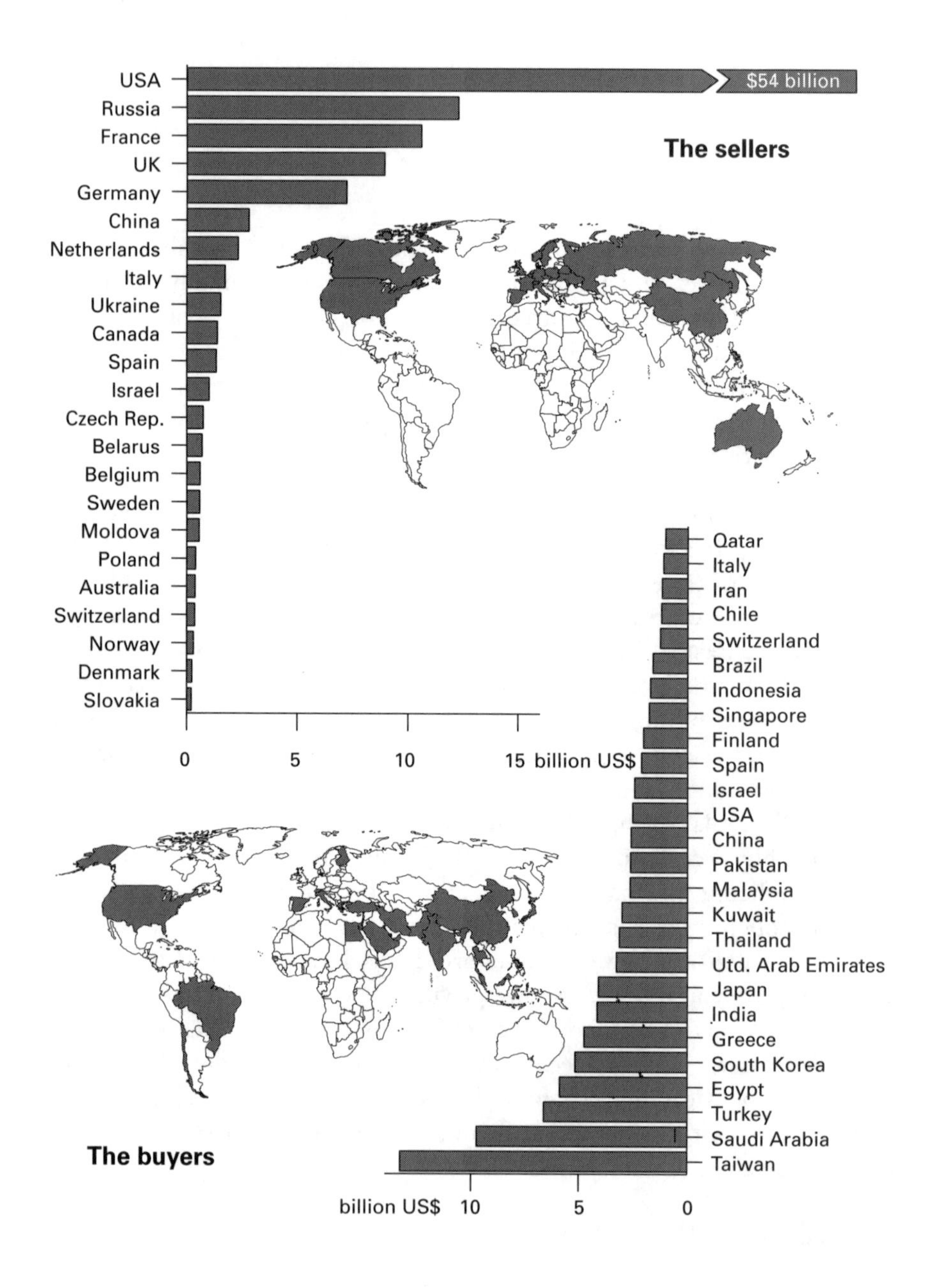

USA
Russia
France
UK
Germany
China
Netherlands
Italy
Ukraine
Canada
Spain
Israel
Czech Rep.
Belarus
Belgium
Sweden
Moldova
Poland
Australia
Switzerland
Norway
Denmark
Slovakia
$54 billion
The sellers
0
5
10
15 billion US$
Qatar
Italy
Iran
Chile
Switzerland
Brazil
Indonesia
Singapore
Finland
Spain
Israel
USA
China
Pakistan
Malaysia
Kuwait
Thailand
Utd. Arab Emirates
Japan
India
Greece
South Korea
Egypt
Turkey
Saudi Arabia
Taiwan
The buyers
billion US$ 10
5
0

● The two bar charts show the principal countries participating in the international sale and purchase of arms. The maps simply indicate these countries.

○ The three types of commodities most traded in the world are believed to be petroleum and its products, addictive drugs and arms. The first striking feature of this map is the overwhelming importance of the USA in the international arms trade. US arms sales amount to nearly $10 per head of the world's population, and about $200 per head of the US population. And the second is that of the first six national arms sellers, five are members of the United Nations Security Council and the other is pressing hard to become one. Arms exports come almost by definition from developed industrialized countries. These are industries which are usually heavily protected by the purchasing policies of national militaries. And they export their surplus production in general to developing countries in the most conflictive areas of the world, something made very evident by the visual comparison of the two maps.

A few countries appear in both lists as major sellers and buyers of arms; they include the USA, China, Italy and Spain. This indicates that there is a certain amount of specialization even in nationally protected arms industries and so deficiencies are made up by imports. Despite the fifty years of military cooperation in NATO, however, not to mention the spread of free-trade ideologies, developed countries still in general regard dependency on imported arms as a source of national weakness and arms industries are heavily protected.

■ SIPRI 2000.

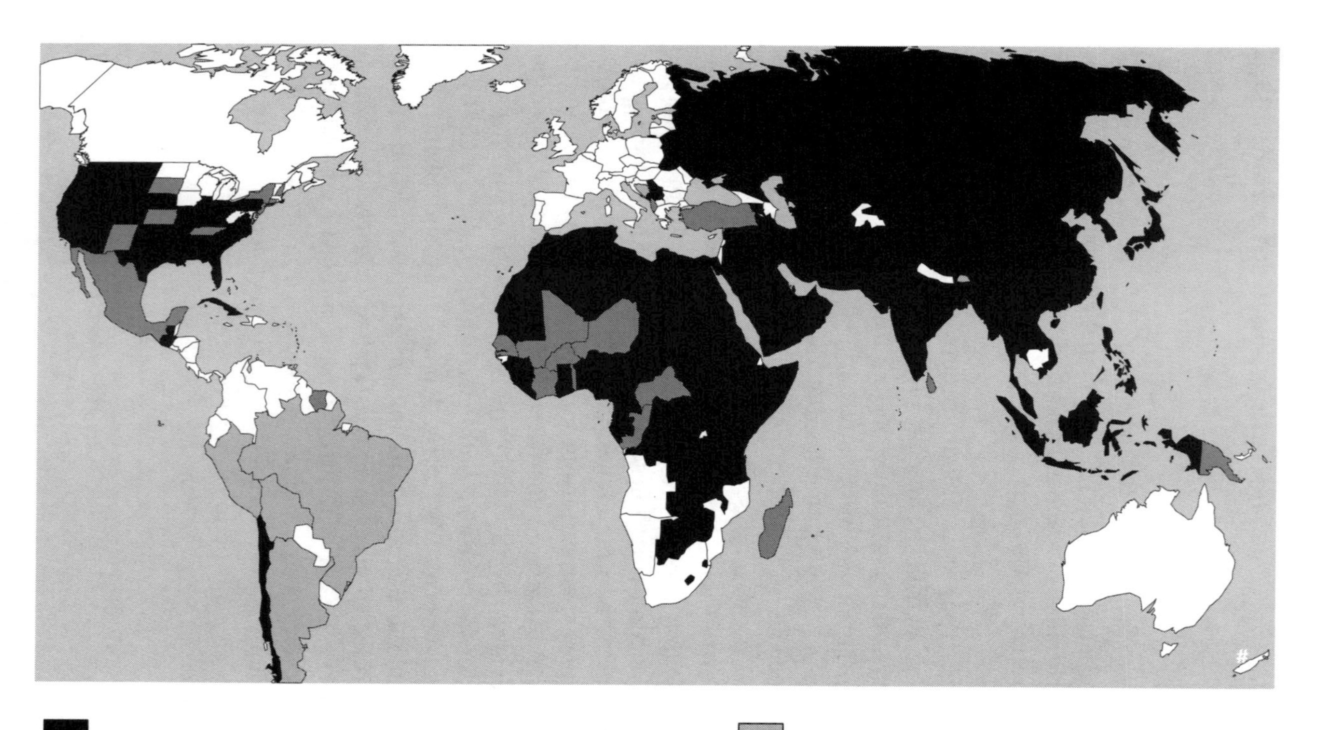

Retentionist (87 countries)
Abolitionist in practice (no executions for ten years) (22 countries)
Abolitionist for ordinary crimes only (13 countries)
Abolitionist for all crimes (73 countries)

● The map shows the state of the death penalty in all the countries of the world according to categories defined by Amnesty International. 'In practice' abolition of the death penalty means that it remains in force but has not been implemented for at least ten years. In the USA the death penalty, after a brief period of abolition, was permitted by a Supreme Court decision in 1976, since when most states have reintroduced it. The map shows the states of the USA according to the same criteria as other countries.

○ The legitimacy of a state's assuming the legal right to kill its own citizens is one of the most controversial questions in political philosophy. Out of many other differences between countries regarding civil rights, this one has been selected for inclusion here because it illustrates that there is very little correlation between a country's economic level and its position on such crucial questions. The debate about human and civil rights which followed the defeat of fascist regimes led to the abolition of the death penalty in a number of countries. Abolition is a condition of membership of the European Union. The death penalty has also been abolished in Australia, Canada, a few Latin American countries, southern Africa and a handful of Asian countries. In most of Latin America and a number of African and Asian countries it exists but has not recently been invoked. In the most developed country of all, the USA, the death penalty exists and is increasingly invoked, although at the start of the twenty-first century it was once more being questioned publicly by important politicians. It is a central plank in the philosophy of the US state and yet reveals a deep ideological rift among the countries often regarded as ideologically homogeneous.

■ Amnesty International 2000; Death Penalty Information Center 2000.

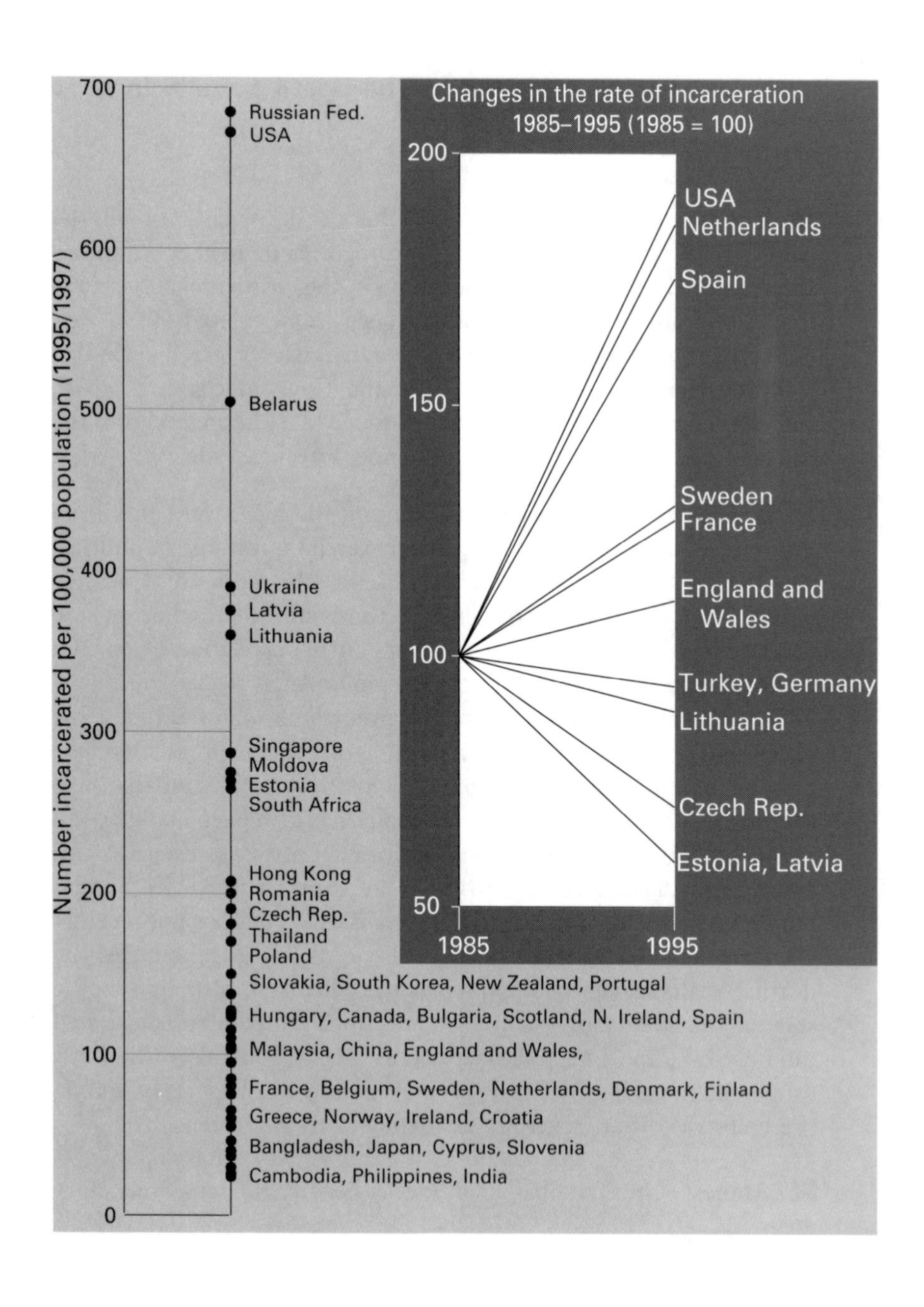

Number incarcerated per 100,000 population (1995/1997)
700
600
500
400
300
200
100
0
Russian Fed.
USA
Belarus
Ukraine
Latvia
Lithuania
Singapore
Moldova
Estonia
South Africa
Hong Kong
Romania
Czech Rep.
Thailand
Poland
Slovakia, South Korea, New Zealand, Portugal
Hungary, Canada, Bulgaria, Scotland, N. Ireland, Spain
Malaysia, China, England and Wales,
France, Belgium, Sweden, Netherlands, Denmark, Finland
Greece, Norway, Ireland, Croatia
Bangladesh, Japan, Cyprus, Slovenia
Cambodia, Philippines, India
Changes in the rate of incarceration
1985–1995 (1985 = 100)
200
150
100
50
1985
1995
USA
Netherlands
Spain
Sweden
France
England and
Wales
Turkey, Germany
Lithuania
Czech Rep.
Estonia, Latvia

● The vertical graph shows the level of incarceration in those countries for which comparative information exists.The figures for the USA and Russia are for 1997, the rest for 1995. The inset graph shows changes in this level in a smaller number of countries between 1985 and 1995.

○ The extent to which a state feels the need to deprive its citizens of their physical freedom is a centrally important element in the definition of the nature of that state, the society which exists within its borders and the relation between the two. The country whose leaders claim most for the freedom of its political system is one of the two which most restricts the physical freedom of its own citizens. Its only rival is its erstwhile enemy in the fifty-year-long Cold War. Another curiously contradictory case is Spain, where during the passage from the Franco dictatorship to a modern democracy the rate of incarceration nearly doubled, as it did in the same period in the Netherlands, generally seen as one of the most tolerant of modern industrialized societies. Of course, the increases in incarcerated people is not, at least overtly, an increase in political prisoners but in people convicted of criminal offences. To be more specific, in 'Western' countries the increase in incarceration has been very closely associated with drug-related offences.

Why is incarceration so much more reverted to by the state in the USA and Russia than in all other countries? Why can India, sometimes regarded as an extremely conflictive society, exist with only one-fiftieth of the proportion of its population in jail as the USA and Russia? Is it because of differences in the crime rate, detection and policing methods or the degree of repression? The dramatic nature of these still little-known figures make those questions very important ones.

■ The Sentencing Project 1999; Mauer 1997.

Composition of United States prison population, 1997

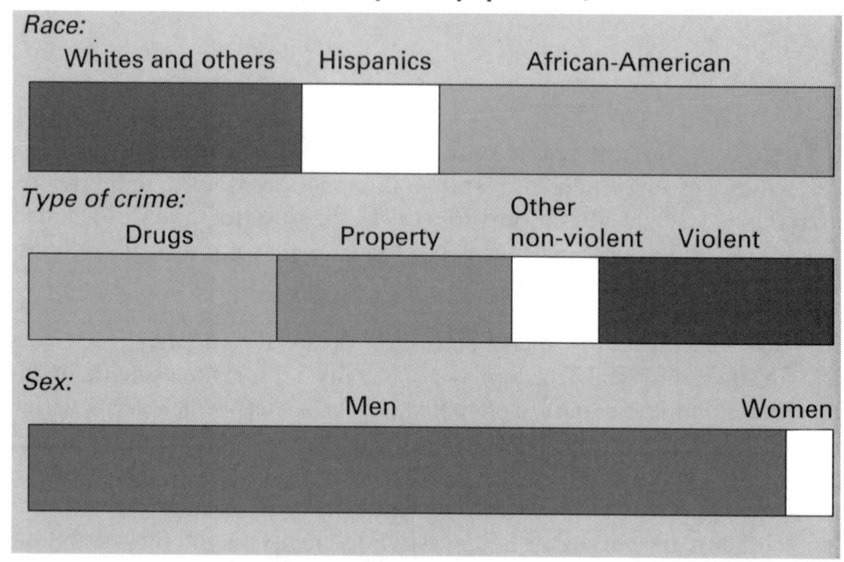

United States incarceration rates (percent of population, 1997)

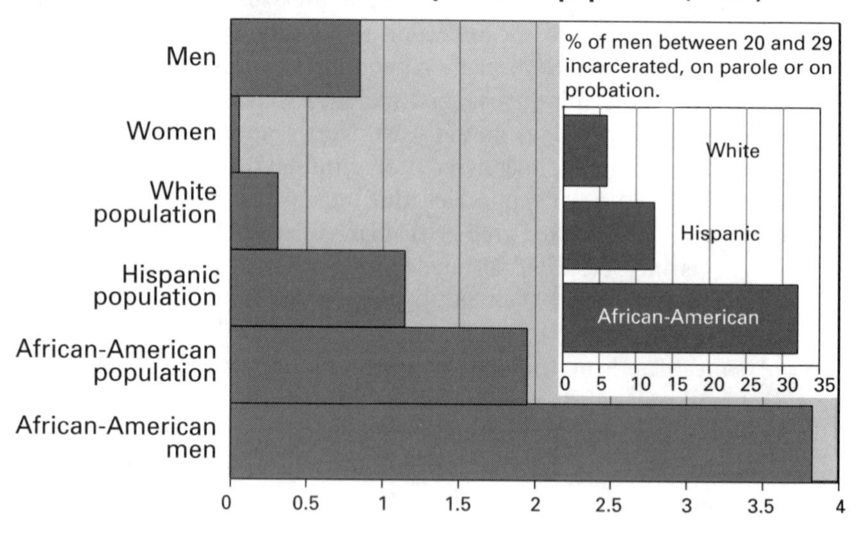

● The upper chart shows the division of the prison population of the USA according to race, type of crime and sex. And the lower shows the incarceration rates in percentage of the population for different segments of the population. The inset graph shows differences in incarceration and judicial supervision of young men according to ethnic group.

○ In late 1999 the doors of a US prison opened to admit the country's 2 millionth prisoner. But the US prison population is very far from being a social cross-section of the country. Some 95 per cent of prisoners are men. Nearly one-third are in prison for crimes involving violence, a similar number for drugs offences and a similar number for property offences (thefts). In these respects the USA is not strikingly different from many other countries. But in another respect it almost certainly is. One-half of prisoners are African-Americans, who form only about 12 per cent of the population; and 17 per cent are 'Hispanics', who also form about 12 per cent of the population. If any other country imprisoned members of a generally poorer ethnic minority population at over seven times the rate at which the ethnic majority population was imprisoned it is hard to imagine that Washington would not have something to say about it, including the imposition of punitive economic sanctions for the violation of human rights.

At each level of detail something more is revealed about what the high US incarceration figures mean. In the inset graph it can be seen that the group which suffers the greatest impact of the high incarceration regime is young African-American men. About one-third of people in this group are either in jail or under some other kind of judicial control such as parole or probation.

■ The Sentencing Project 1999; Mauer 1997.

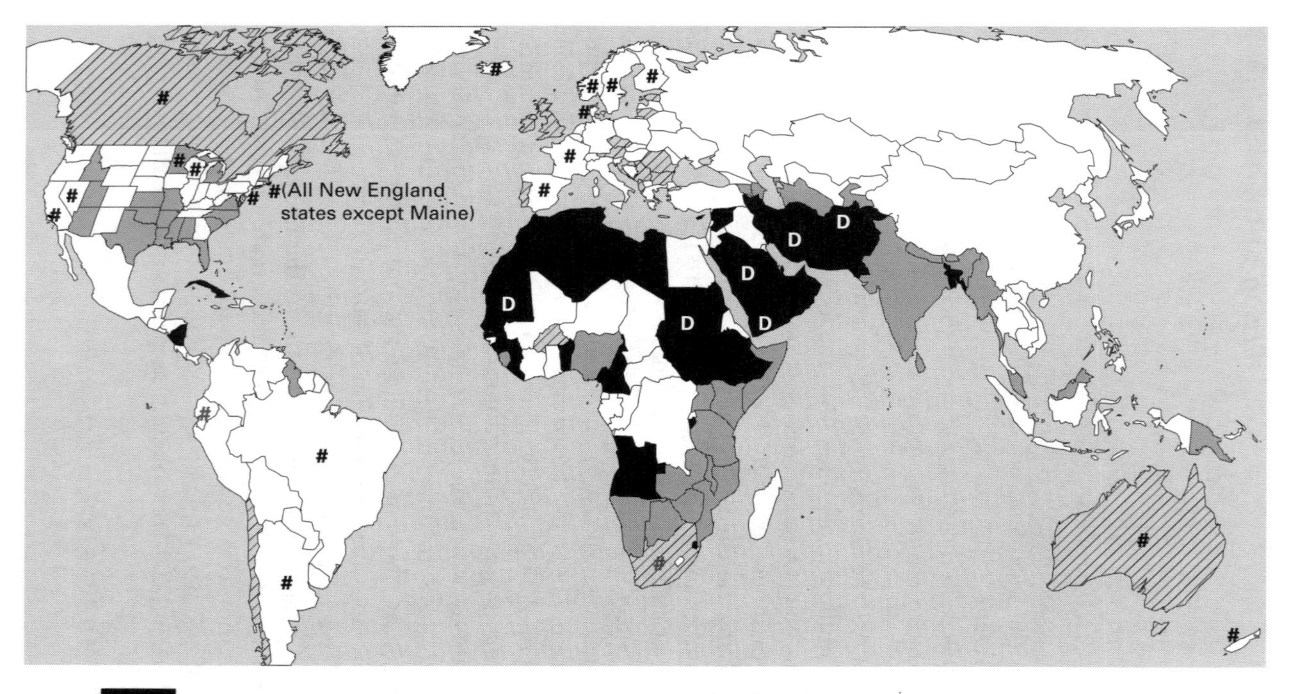
#(All New England
states except Maine)
Gay male and lesbian sex illegal
Gay male sex illegal
Anti-gay and lesbian legal discrimination
# = legal protection against discrimination
D = death penalty for gay sex

● The map shows various aspects of the legal treatment of homosexuality. The actual degree of repression and acceptance may differ greatly from the legal provisions.

○ Civil and human rights could be measured by the treatment of numerous groups. Since there is not enough space for that here, the treatment of homosexuality has been included, in part as an example of the many international differences in social tolerance, but also because sexual behaviour regarded by some as deviant, homosexuality in particular, is often the most 'difficult' of human rights for a society to accept. It is therefore a kind of test case, often persecuted by the right and forgotten by the left. It is also included here because of its personal effect on the author of this book.

The map is in some ways surprising. An enormous variety of legal treatments of homosexuality are observable, varying between the existence of legal protection against discrimination on grounds of sexual preference to an obligatory death sentence. Legal treatment is not necessarily reflected in social treatment. In Brazil, for instance, there is no legal constraint – there is even an anti-discrimination law – but there is also a high level of anti-gay violence.

In the USA and in Australia the legal situation differs between states. In the USA there is almost as much variety as there is in the world as a whole. The most repressive legal regimes are to be found where conservative forms of religion are strong. But religion does not determine everything, as is shown by the fact that countries traditionally dominated by the Catholic Church, as opposed to homosexuality as conservative Islam, have in many cases eliminated anti-gay and lesbian laws.

■ ILGA 2000; NLGTF 1999.

Inequality and history X

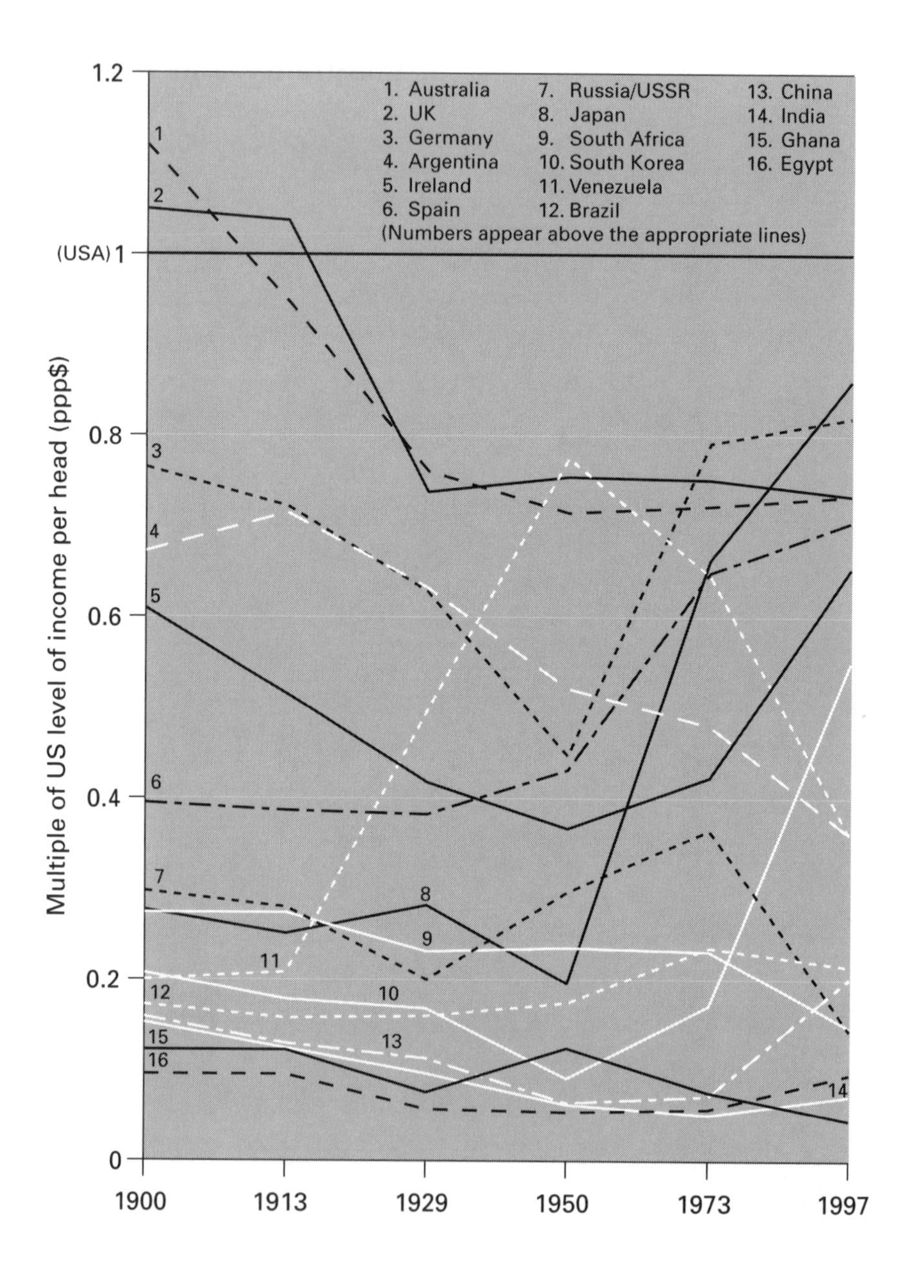
1. Australia
2. UK
3. Germany
4. Argentina
5. Ireland
6. Spain
7. Russia/USSR
8. Japan
9. South Africa
10. South Korea
11. Venezuela
12. Brazil
13. China
14. India
15. Ghana
16. Egypt
(Numbers appear above the appropriate lines)
Multiple of US level of income per head (ppp$)
(USA) 1
1.2
0.8
0.6
0.4
0.2
0
1900
1913
1929
1950
1973
1997

The evolution of relative incomes, 1820–1997

● The graph is based on estimates of real **national income** per head converted using **purchasing power parity** so that they are in principle comparable between countries and between dates. Data for a representative sample of countries are shown each as a multiple of the figure for the USA, which therefore stays constant at 1.

○ The figures summarized in this graph are probably the best historical comparisons of the level of international income per head that will ever be available. If they are accurate they show a number of very important facts, surprising and unsurprising. In the first half of the last century a long process of divergence between the USA and the other developed countries, which had begun in 1850, continued. By 1950 the USA was by far the richest country in terms of income per head. The immense costs of World War II, especially in Europe and Japan, left the USA in 1950 with its largest ever advantage over other countries. Since then countries of western and southern Europe have converged with the USA, although they have failed to reach its income level. Japan's convergence was the latest and has gone furthest. Meanwhile during this long history a few countries, for instance Argentina, Venezuela and Russia, achieved a period of temporary convergence only to fall back again. During the last fifty years Latin American countries converged with each other but diverged from the USA.

Almost throughout the period the relative income levels of Asian and African countries fell continuously. At the end of the period, however, China's income per head began to converge with that of the USA, as did that of India, though much more slowly. South Korea's rose much more quickly. Most African countries have not yet broken the downward relative trend.

■ Maddison 1995; World Bank 1999a.

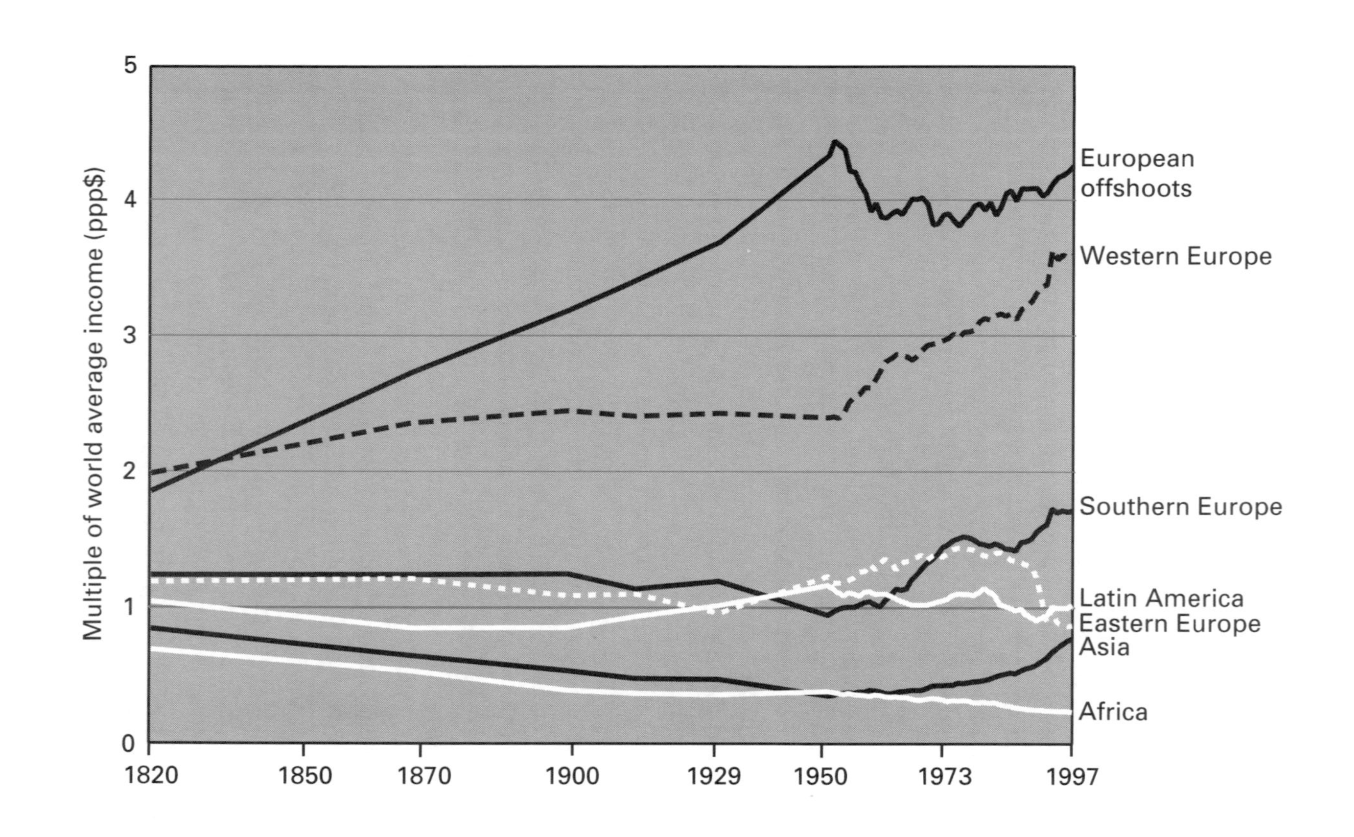
Multiple of world average income (ppp$)
0
1
2
3
4
5
1820
1850
1870
1900
1929
1950
1973
1997
European offshoots
Western Europe
Southern Europe
Latin America
Eastern Europe
Asia
Africa

The long-term polarization of world income, 1820–1997

● This graph is another way of seeing the same data that appear in the previous one. Here countries have been grouped into types or geographical regions and the standard of comparison is the average world level of income which therefore appears as 1 in the graph. 'European Offshoots' comprises the USA, Canada, Australia and New Zealand. The category 'Southern Europe' includes Turkey. The years chosen are benchmark years for the world economy but after 1950 the data are plotted annually.

○ This graph makes even clearer the existence of a tendency towards divergence between rich and poor countries in their income levels during 200 years in which industrial capitalism has come to dominate the world economy. But in detail there are a number of partial divergences and convergences within this general pattern. Eastern Europe rose relatively until about 1973 and has fallen relatively since then. Latin America fell until 1870, rose slowly up to 1950 and has fallen since 1980. Southern Europe has closed some of the gap with the richer countries which opened up between 1820 and 1950. Asia fell continuously until 1973 and then began to close the gap; it has now returned approximately to its relative level of 1820. Africa continues to show long-term relative decline.

According to these figures the ratio between the richest and poorest regions of the world in 1820 was less than 3 to 1; it has now risen to about 16 to 1. There is no doubt that the extremes have drawn further apart. But between the extremes there have been and continue to be complex partial long-term and short-term processes of convergence and divergence (see Introduction).

■ Maddison 1995; World Bank 1999a.

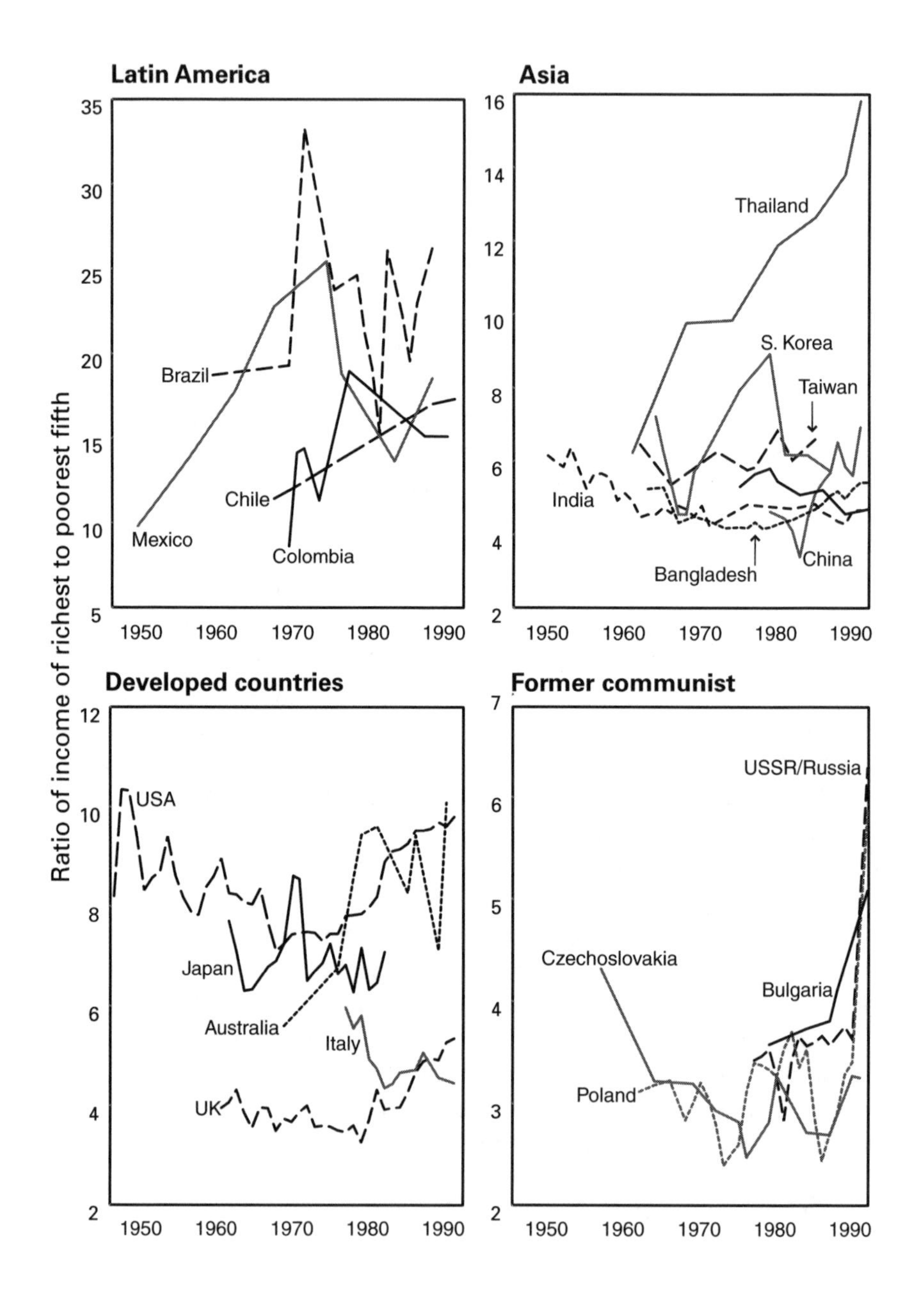

Latin America
Asia
Developed countries
Former communist
Ratio of income of richest to poorest fifth
35
30
25
20
15
10
5
1950
1960
1970
1980
1990
Brazil
Mexico
Chile
Colombia
16
14
12
10
8
6
4
2
Thailand
S. Korea
Taiwan
India
Bangladesh
China
12
10
8
6
4
2
USA
Japan
Australia
Italy
UK
7
6
5
4
3
2
USSR/Russia
Czechoslovakia
Bulgaria
Poland

● The four graphs show for the indicated countries the evolution of the ratio between the income of the richest and the poorest 20 per cent of the population. This set of data does not go much beyond 1990.

○ The overall level of inequality in the world is determined not merely by the level of inequality between nations, but also by the evolution of inequality within nations. Historical data on the internal national distribution of income are much scarcer and even less reliable than those for the average national income per head. Roughly comparable figures for various parts of the period between 1950 and 1995 show the complex picture in these graphs.

In Latin America inequality is very high and in some countries has moved fairly erratically. There has been a general tendency during the period for inequality to grow. In Asia national inequality is typically much lower and in some important countries has, according to these figures, declined. But in a few inequality has grown: especially in China since 1980 and in Thailand since 1960. In the latter country inequality has now reached Latin American levels. In the developed countries there is considerable variation between countries and periods. In Italy and Japan inequality has been reduced. In the USA inequality also fell until about 1965 since when it has risen sharply (❮ 16). The UK shows a similar pattern, with the turning point in 1979 (❮ 15). In Australia since the 1960s there has been a sharp increase in inequality. Finally, in the formerly communist countries, which have always shown the lowest figures for inequality, the transition to capitalism has been marked (except apparently in the case of the former Czechoslovakia) by an exceptionally sharp increase in inequality.

■ Deininger and Squire 1996.

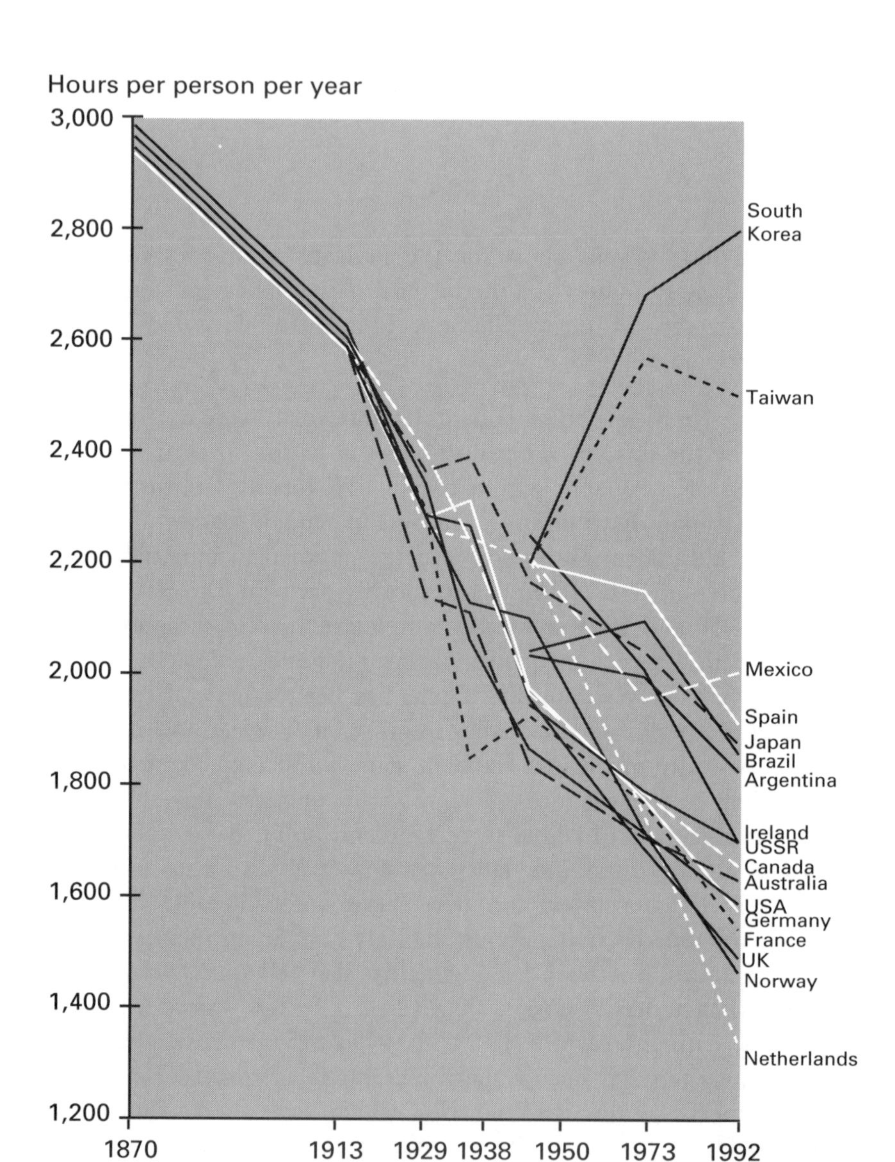

Hours per person per year
3,000
2,800
2,600
2,400
2,200
2,000
1,800
1,600
1,400
1,200
1870
1913
1929
1938
1950
1973
1992
South Korea
Taiwan
Mexico
Spain
Japan
Brazil
Argentina
Ireland
USSR
Canada
Australia
USA
Germany
France
UK
Norway
Netherlands

Contrasting movements in the hours of work, 1870–1992

● The lines indicate the average hours worked by production workers in the countries indicated between 1870 and 1992.

○ The wide differences in the hours of work between different countries is seldom taken into account in comparing international indicators of development although it has a profound effect on living standards (‹ 6, 7).

In 1870 it was typical for workers to work a 3,000-hour year, equivalent to a 60-hour week with no holiday. Trade unions and workers' movements have generally waged campaigns for hours to be reduced to more humanly tolerable levels. The world now celebrates May Day, though few know that it is to commemorate a group of martyrs to the cause of the 8-hour day in the United States. This graph, based on rather limited data, shows how hours worked have fallen erratically but greatly since then. In most developed countries in the 1990s they varied between 1,400 and 1,600 hours (‹ 6). In Latin America, Japan and Spain they remain somewhat higher.

The most striking facts in this graph are that, in the two countries most often cited as the great success stories of modern development, hours worked – which were in the upper part of the international spectrum in 1950 – have, instead of falling as in all other countries, risen sharply, in the case of South Korea to levels not much below Europe in 1870. If these figures are correct (and they are not completely consistent with other estimates) they put a rather different slant on the impressive economic growth of South Korea and Taiwan. They suggest that a major part of it was based not on a miracle blossoming of labour productivity so much as an all-too-mundane increase in work hours and intensity.

■ Maddison 1995.

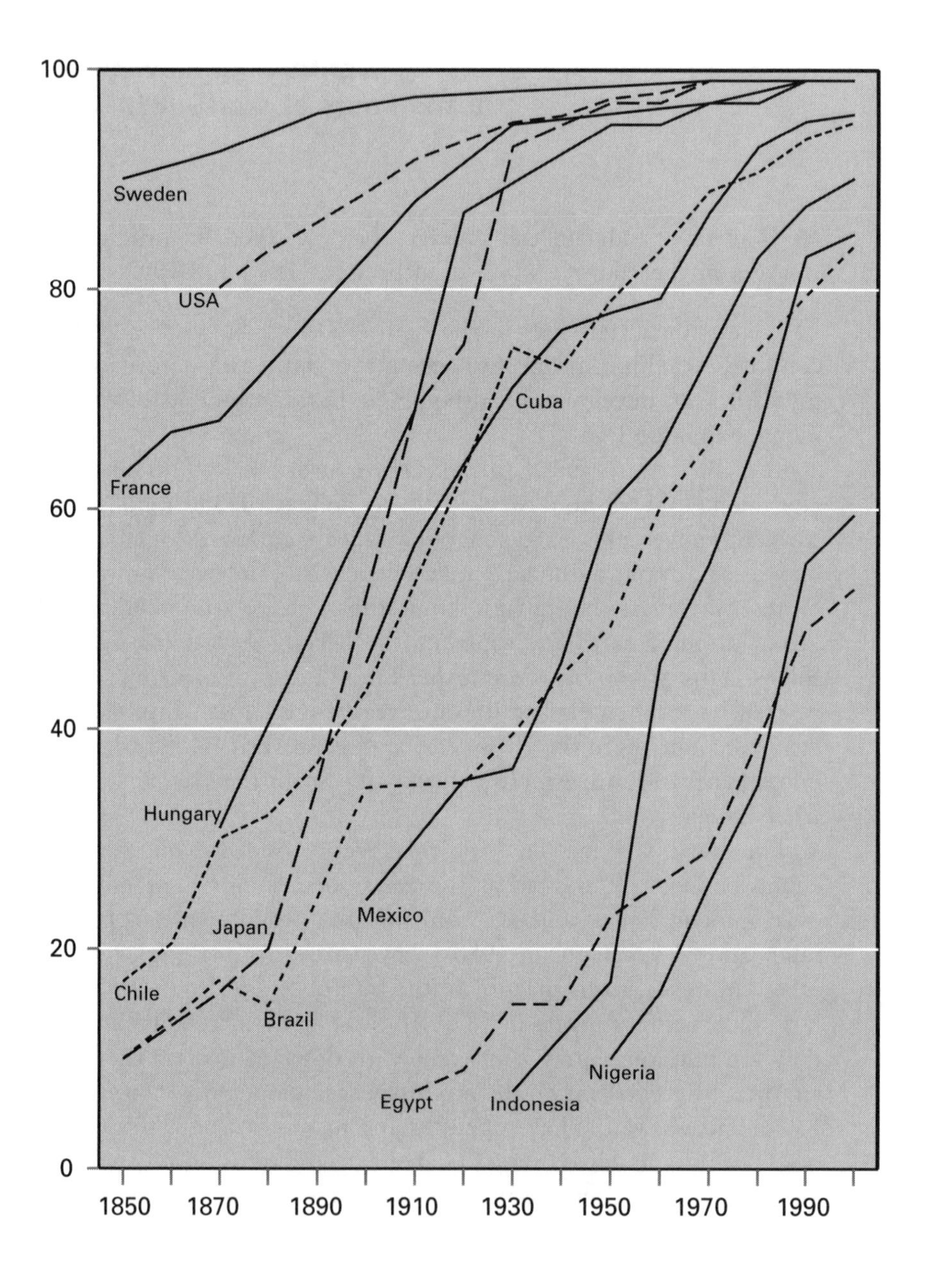

100
80
60
40
20
0
1850
1870
1890
1910
1930
1950
1970
1990
Sweden
USA
France
Cuba
Hungary
Japan
Mexico
Chile
Brazil
Egypt
Indonesia
Nigeria

The growth of literacy, 1850–1995

● The lines show the evolution of levels of adult literacy over the period since 1850.

○ Literacy is an indicator of differences in levels of development which has been increasingly used in recent years as a complement to the level of income. Education both contributes to the standard of living by providing access to culture and contributes to further development by creating human capital.

The figures show a clear and in some places quite rapid convergence, a very different picture from that given by income and some other indicators (‹ 114, 115). Since literacy is one of the indicators included in the Index of Human Development it helps to impart to that index a tendency towards convergence between countries (›122). Statistically this convergence is inevitable since the indicator has a natural limit of 100 per cent. So all progress towards this limit appears as convergence.

Against the idea that this represents a real convergence, some reservations could be made. One is that literacy itself is not a very good measure because it is defined differently in different places and the criteria are sometimes minimal. Second, literacy itself can mean little without also taking into account whether, once it is attained, it can be put to use. In other words, literacy needs complements, including the opportunity for education, access to written matter and to writing materials. Third, it is sometimes complained that literacy is a rather Eurocentric criterion of development and that culture can also get destroyed by formal education.

Cynically one might say that the final objection is almost always made by literate people who would hate to lose their literacy. And the other objections only amount to saying that literacy is not an end but a means.

■ Maddison 1995.

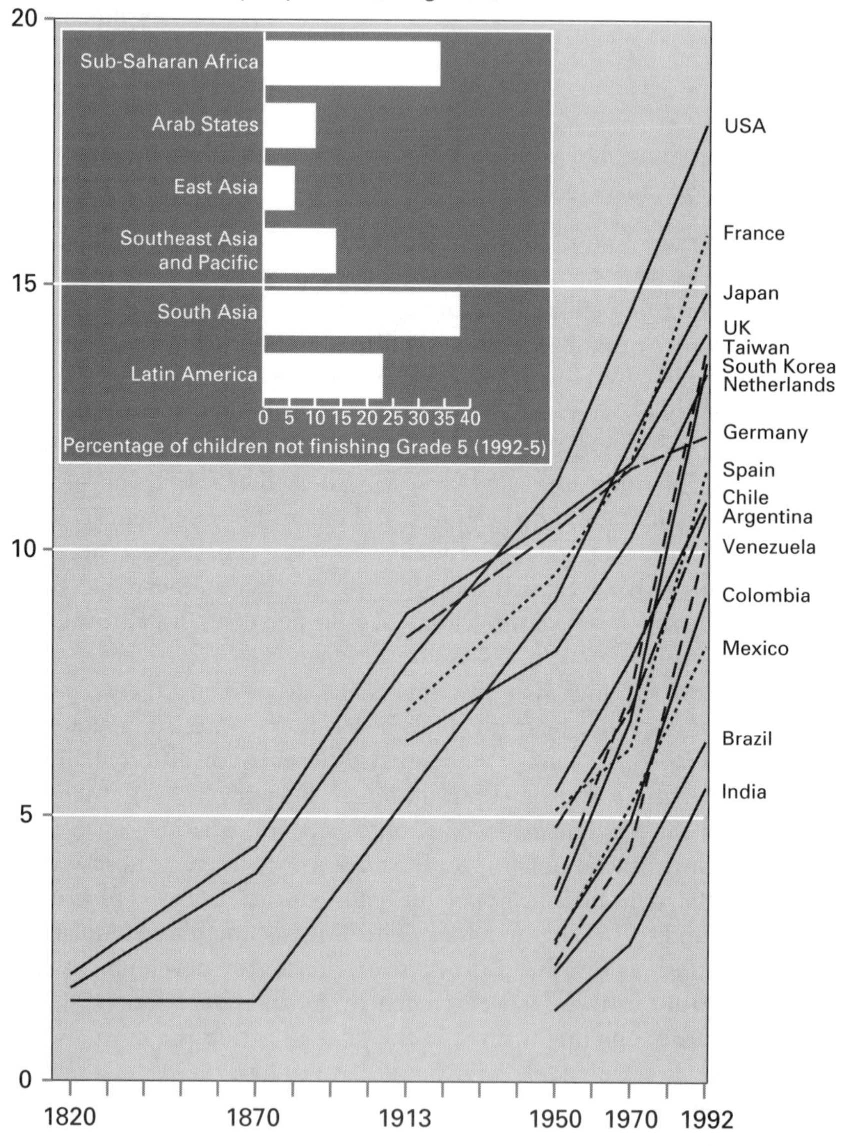

Years of education per person (weighted)
20
15
10
5
0
1820
1870
1913
1950
1970
1992
Sub-Saharan Africa
Arab States
East Asia
Southeast Asia and Pacific
South Asia
Latin America
0 5 10 15 20 25 30 35 40
Percentage of children not finishing Grade 5 (1992-5)
USA
France
Japan
UK
Taiwan
South Korea
Netherlands
Germany
Spain
Chile
Argentina
Venezuela
Colombia
Mexico
Brazil
India

● The lines represent a weighted average of years of education per person in all kinds of school. The inset chart shows another measure of educational difference, the percentage of children who fail to complete primary school.

○ Years of education is not a variable which has a natural limit like literacy. But there may be an effective limit when nearly all those who wish to receive a college education. The countries at the top of this list continue to show improvements because they are still well within this limit. The quantitative range between countries according to this criterion is much wider than for variables such as life expectancy or literacy, even if not as wide as differences in income levels. There has been considerable convergence in this indicator except for the poorest countries (mostly in Africa and South Asia), which continue to lag far behind. In particular the East Asian countries which have had very fast economic growth rates (Taiwan and South Korea) have reached the same levels as the most developed countries according to this variable, a fact which, unlike hours of work (‹ 117), does suggest that their development is much more than simply getting people to work more.

The other measure of educational attainment shown here is the percentage of children who never complete primary school. About 40 per cent of children are in this category in Sub-Saharan Africa and in South Asia. But the level is high also in Latin America (nearly 25 per cent) and parts of Southeast Asia (over 15 per cent in aggregate). As we saw earlier (‹ 58), a considerable amount of the deficiencies observed in any educational variable are the result of sexual bias, since girls still in most places have lower access to education than boys.

■ Maddison 1995; UNDP 1999.

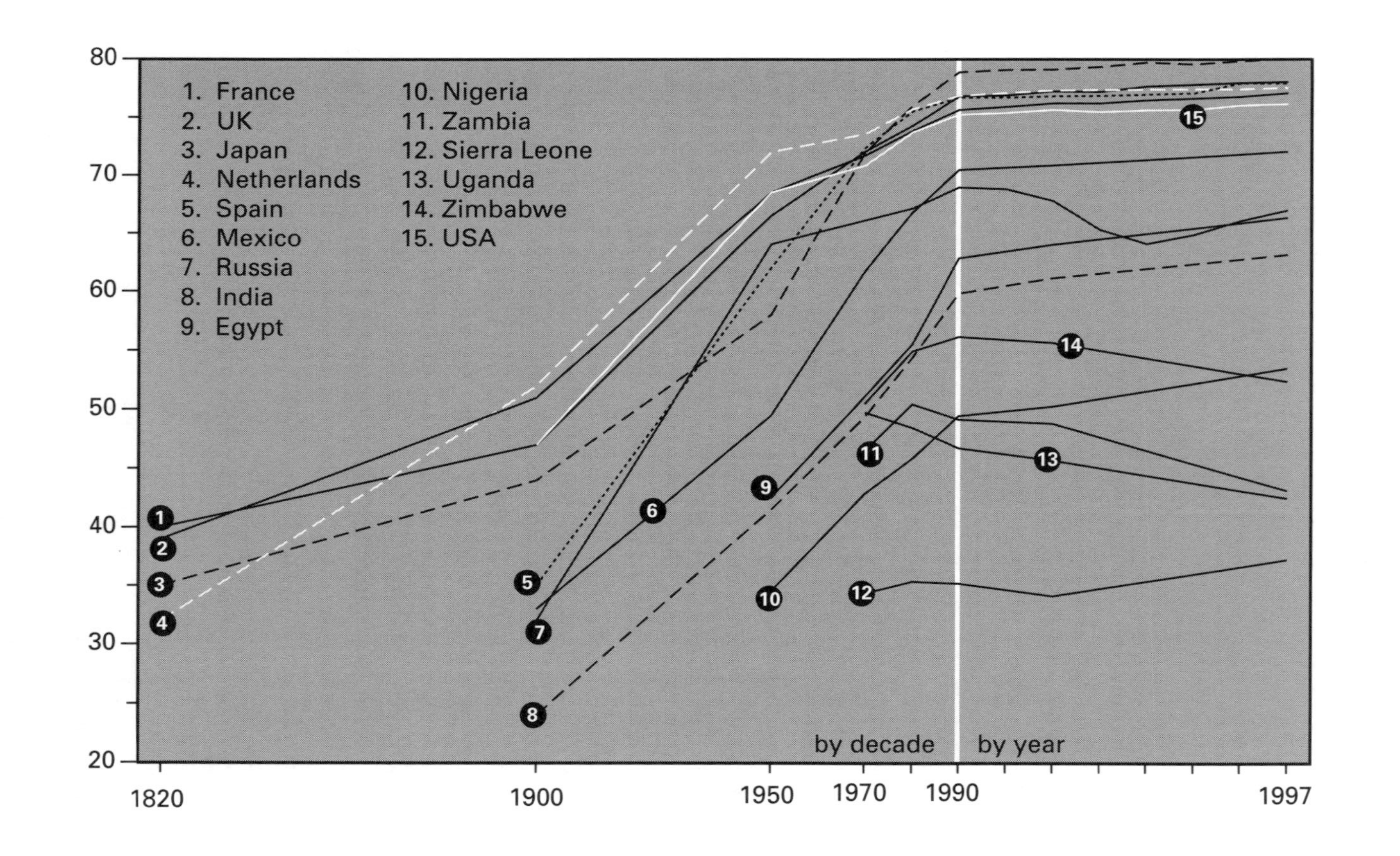

1. France
2. UK
3. Japan
4. Netherlands
5. Spain
6. Mexico
7. Russia
8. India
9. Egypt
10. Nigeria
11. Zambia
12. Sierra Leone
13. Uganda
14. Zimbabwe
15. USA
80
70
60
50
40
30
20
1820
1900
1950
1970
1990
1997
by decade
by year

Rises and falls of life expectancy, 1820–1997

● The lines show the evolution of life expectancy at birth (for the whole population) by decade from 1820 to 1990 and then annually up to 1997. The data are given for a representative sample of countries.

○ Life expectancy at birth, another of the indicators that make up the Human Development Index, has in effect a biological limit. It is hard to imagine it going much beyond 85 in the foreseeable future. It has reached almost 80 years in a few countries such as Japan and Western Europe. It is to be expected therefore that the indicator will show strong convergence. What in fact is seen here is:

- a strong increase in life expectancy during the twentieth century in all the developed countries, due especially to the reduction of fatal infectious diseases, and other causes of infant and maternal mortality. It is the combined result of greater general health knowledge, medical discoveries and improved public hygiene (‹ 23, 37);
- a strong convergence of many developing countries after 1950 as they too applied the same improvements;
- a recent fall in life expectancy in a few countries. Some of these are African countries and the cause is almost entirely the AIDS epidemic. But life expectancy has also deteriorated in Russia and other parts of the former USSR – caused, analysts have concluded, by the social trauma, especially for middle-aged men, of the transition from the relatively stable full-employment society of communism.

■ Maddison 1995; World Bank 1999a.

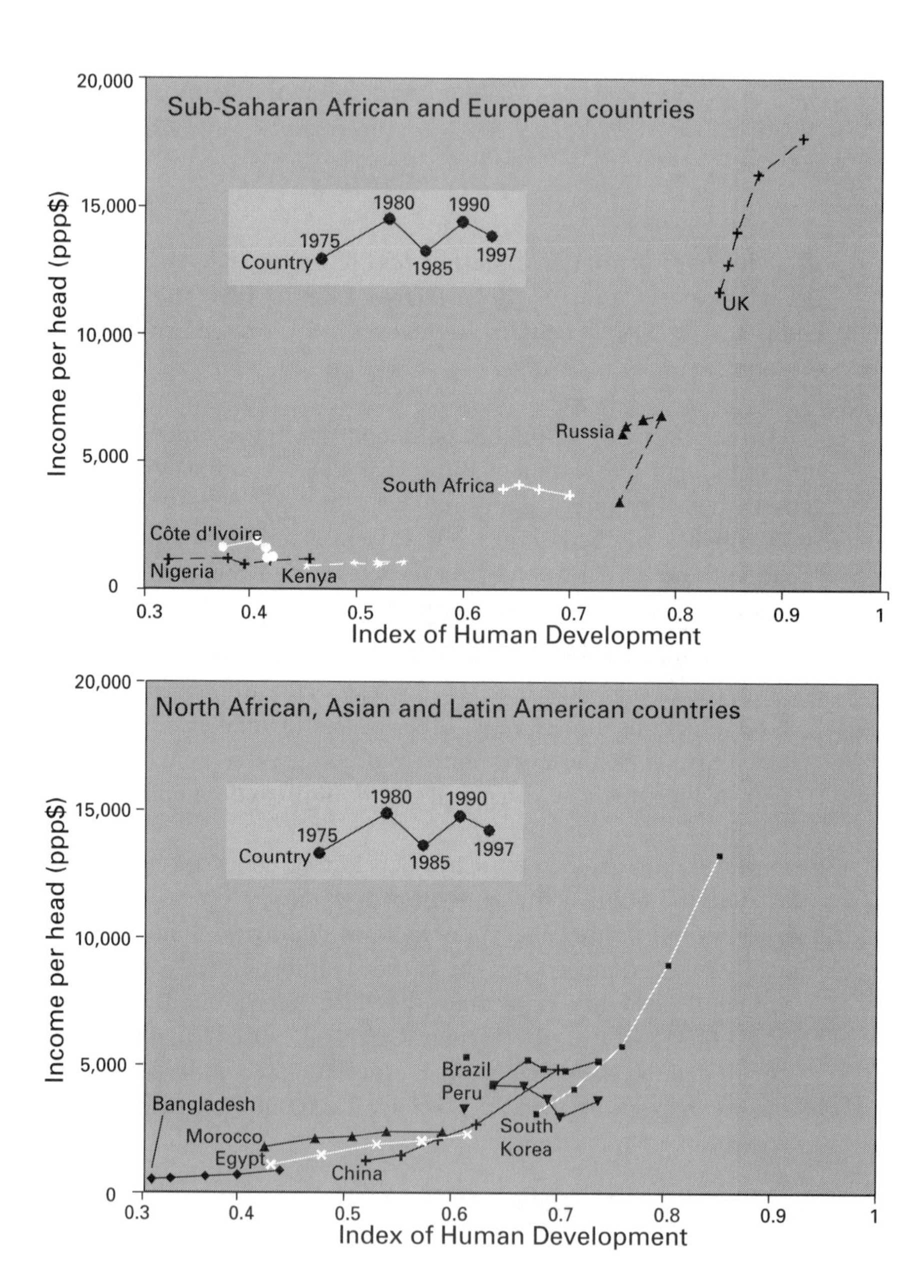
Sub-Saharan African and European countries
Income per head (ppp$)
Index of Human Development
Country
1975
1980
1985
1990
1997
UK
Russia
South Africa
Côte d'Ivoire
Nigeria
Kenya
North African, Asian and Latin American countries
Brazil
Peru
Bangladesh
Morocco
Egypt
China
South
Korea

The combined evolution of income and human development, 1960–1997

● The upper and lower graphs are identical. They show for a selected group of countries the combined evolution since 1975 of the two most widely used measures of international differences in the level of development: income per head and the relatively new Human Development Index (‹ 21).

○ These graphs show the great variety of recent experience in relation to the two most common measures of development. Almost all combinations are visible here in different countries:

- strong growth of income and HDI: South Korea, UK;
- strong growth of HDI, little growth of income: Bangladesh, Nigeria, Kenya, Peru, Brazil;
- moderate HDI growth, decline of income: Peru;
- slow growth of HDI, decline of income: South Africa, Côte d'Ivoire;
- decline of HDI and decline of income: Russia.

Such a variety of experiences can be used to support the hypotheses of those who argue that there is a strong relationship between the two concepts of development, either because economic growth permits human development or because human development (better education and health) stimulates growth. But there are examples which do not follow this pattern of combined increase and which might be used to support another idea, often defended by the UNDP, that the two are to some extent independent and that with enlightened government policies it is possible to have human development independently of the growth of income. The fact that the evidence is contradictory suggests that there are many things hidden behind these facts.

UNDP 1999.

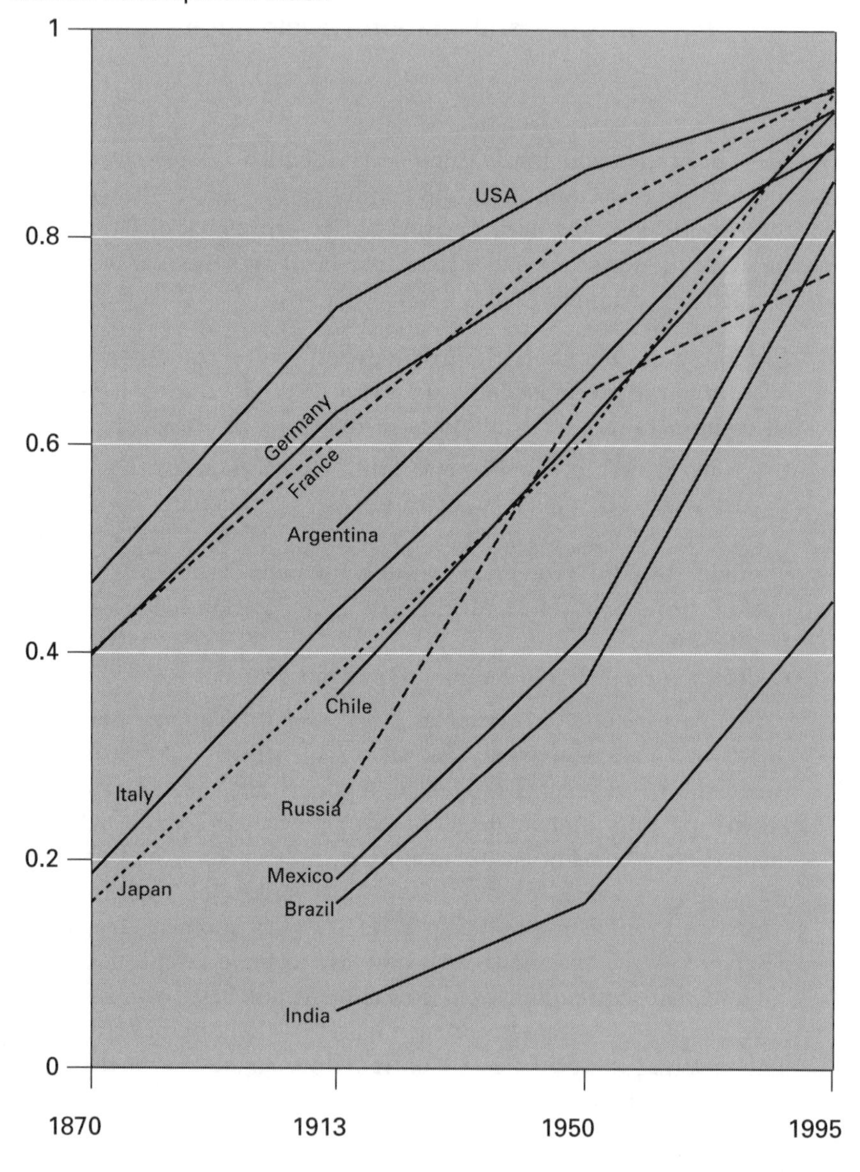

Human Development Index
1
0.8
0.6
0.4
0.2
0
1870
1913
1950
1995
USA
Germany
France
Argentina
Chile
Italy
Russia
Japan
Mexico
Brazil
India

The convergence of the Human Development Index over time, 1875–1995

● The graph shows the levels of the Human Development Index since 1875 for a limited number of countries.

○ The UNDP has calculated the value of its Human Development Index for many countries since 1960. More recently a much longer term estimate of HDI levels has been made for a limited number of countries and the results have aroused considerable interest.

The graph shows that, in contrast to increasing international divergence in levels of income per head, the Human Development Index has shown a strong tendency towards international convergence over the long period. It is ironic that the UNDP, which ten years ago invented the concept and index of human development, and which insisted on the use of **purchasing power parity** methodology in the comparing income levels, now habitually describes the world in terms of the relative level of national income compared using the former methodology of exchange rates. The reason is that those figures show more international divergence, which is convenient for the UNDP's message in favour of government intervention and redistribution. It is equally ironic that, after ten years without even so much as hinting that it knew that the HDI existed, the IMF has recently emphasized the historic international convergence shown by the HDI. This is because it suits *their* message – that in general only competitive capitalism can develop and, implicitly, equalize the world.

These ironies are reminders that statistics are often used in order to support some hidden ideology and that statistical opportunism is rife and should be guarded against – not by ignoring all the statistics, but by looking at them in detail and critically.

■ Crafts 2000.

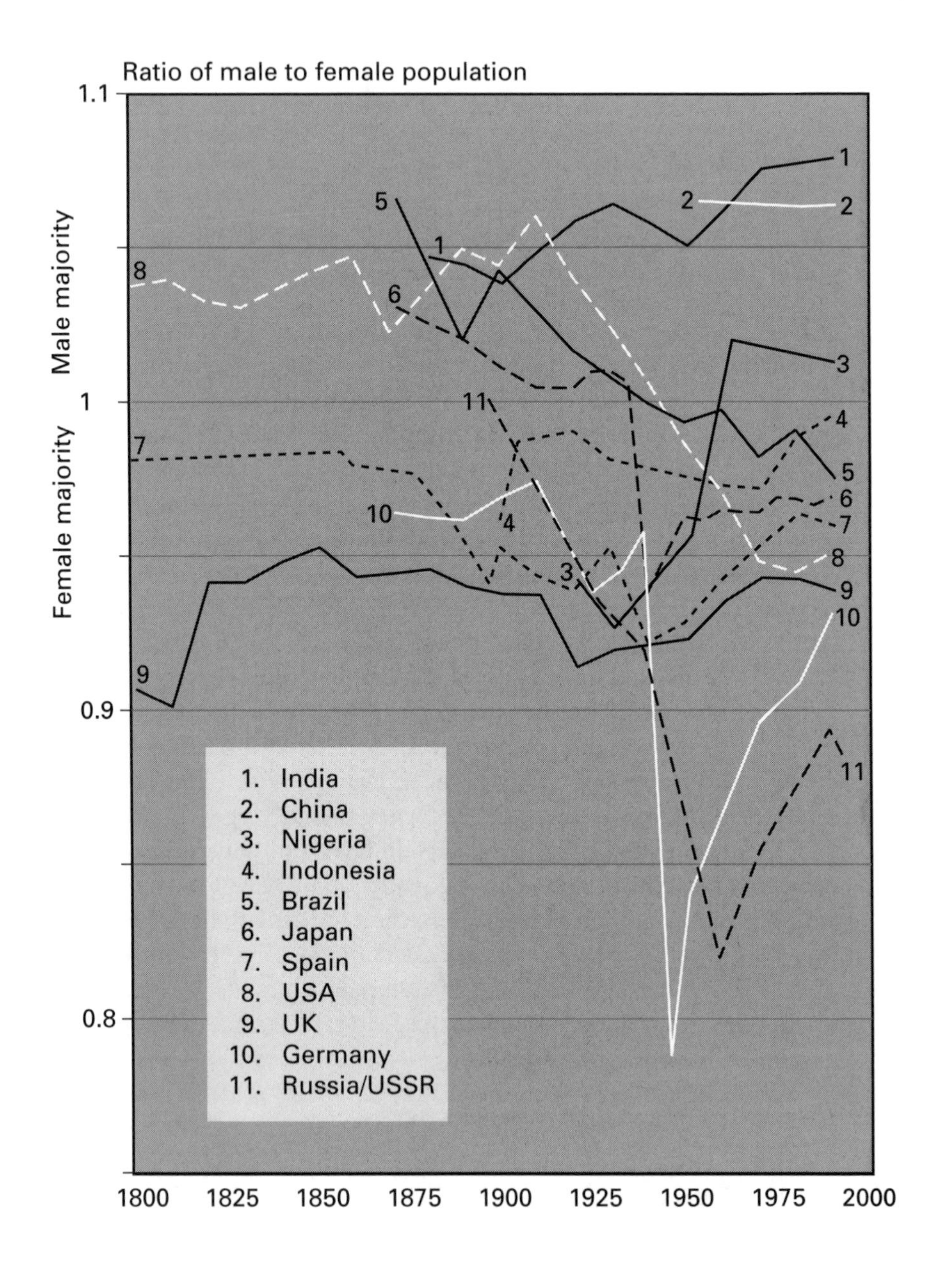
Ratio of male to female population
1.1
1
0.9
0.8
Male majority
Female majority
1800
1825
1850
1875
1900
1925
1950
1975
2000
1. India
2. China
3. Nigeria
4. Indonesia
5. Brazil
6. Japan
7. Spain
8. USA
9. UK
10. Germany
11. Russia/USSR

Historical changes in the population of men and women, 1800–1995

● The lines show the ratio of the male to the female population in a selected sample of countries, starting in 1820. A figure above 1 means there is a male majority and below 1 a female majority.

○ Based on the experience of the author, the fact in this book which will cause most surprise is that there is a majority of men in the world's population (‹ 26–34). This graph makes it easier to see why that is so and why that majority has grown since about 1965, when it first appeared. Most important, of course, are the data for China and India, since between them they contain 37 per cent of the world's population and so have a decisive impact on the world figure. India has had a male majority during the whole of the twentieth century. Since 1950 it has grown due to increasing maltreatment of women and more recently to selective abortions. China has had a steady male majority. In addition to other Asian countries with male majorities, the most populous country in Africa, Nigeria, has changed very rapidly since 1950 from a female- to a male-majority country. At the same time, with the exception of the USA and Brazil, where during the twentieth century a male majority has given way to a female majority, largely as a result of the declining importance of immigrants in the population, other female-majority countries have not shown much change in the ratio of men to women. Since 1945, countries which lost large numbers of men during World War II (in particular Germany and Russia) have shown sharply reduced female majorities as the weight of the wartime generations has slowly been reduced. Looking at all these countries suggests that there is no chance that the trend towards an increasing male majority in the world will be reversed in the foreseeable future.

■ Mitchell 1998a, 1998b, 1998c; World Bank 1999a.

Sources of the data

Acheson, Donald, et al., 1998, *Independent Inquiry into Inequalities in Health, Report* (The Acheson Report), London: The Stationery Office.
Amnesty International, 2000, website against the Death Penalty (http://www.web.amnesty.org/rmp/dplibrary.nsf/current?openview).
AUB (Australian Bureau of Statistics), 1991, National Census, Canberra: ABS (http://abs.gov.au).
CIESIN (Center for International Earth Science Information Network), 2000, *Demographic Data Viewer* (Interactive mapping of 1990 US census data), website data (http://plue.sedac.ciesin.org/plue/ddviewer/).
China Dimensions, website data (http://sedac.ciesin.org/china/).
China State Statistical Bureau, 1995, *China Statistical Yearbook 1995*, Beijing: Government of People's Republic of China.
Crafts, Nicholas, 2000, *Globalization and Growth in the Twentieth Century*, IMF Working Paper, WP/00/44, Washington DC: IMF.
Death Penalty Information Center, 2000, website data (http://www.essential.org/dpic/).
Deininger, Klaus and Lyn Squire, 1996, 'A new data set measuring income inequality', World Bank, September (http://www.worldbank.org/research/growth/dddeisqu.htm).
Demosphere International/Microsoft, 1997, mapstats.xls, database of statistical information included in mapping module of Excel 97.
Economy Watch, 2000, website data (http://www.economywatch.com/database/).
EPI (Economic Policy Institute), 1999, *The State of Working America 1998–99*, Ithaca NY: Cornell University Press.
EPI (Economic Policy Institute), 2000, The Data Zone, recent economic data on website (http://www.epinet.org), Washington DC.
Eurostat, 1998, *GDP in European Union Regions – Estimations for 1994 to 1996*, Brussels: European Union.
FAO (Food and Agriculture Organization of the UN), 1996, *The Sixth World Food Survey*, Rome: FAO.
FAO (Food and Agriculture Organization of the UN), 1997, *The State of Food and Agriculture (SOFA) 1997*, Rome: FAO.

Filmer, Deon 1995, 'Estimating the World at Work', *Background Report* for World Bank World Development Report 1995, Washington DC: World Bank, Office of the Vice President Development Economics.

Filmer, Deon, Elizabeth M. King and Lant Pritchett, 1998, *Gender Disparity in South Asia: Comparisons between and within Countries* (research paper), Washington DC: World Bank.

Financial Times, 1995, supplement on South Africa, 21 November.

Fundacão Instituto Brasileiro de Geografia e Estatistica, 1994, *Anuario Estatistico do Brasil 1994*, Brasilia: Fundacão Instituto Brasileiro de Geografia e Estatistica.

Global Financial Data, 2000, historical economic data on website (http://www.globalfindata.com).

Hacker, Andrew, 1995, *Two Nations: Black and White; Separate, Hostile, Unequal*, New York: Ballantine Books.

IADB (Interamerican Development Bank), 2000, website data (http://www.iadb.org).

ILGA (International Lesbian and Gay Association), 2000, *World Legal Survey*, website data (http://www.ilga.org/Information/Legal_survey/ilga_world_legal_survey%20introduction.htm).

ILO (International Labour Organization), 1999, *Key Indicators of the Labour Market*, Geneva: ILO.

IMF (International Monetary Fund), 1996, *Direction of Trade Statistics Yearbook 1996*, Washington DC: IMF.

IMF (International Monetary Fund), 1999, *Annual Report*, Washington DC: IMF.

India Profile, 2000, website data (http://www.indiaprofile.com).

Inter-American Dialog, 2000, website data (http://www.iadialog.org/immigraf.htm).

Maddison, Angus, 1995, *Monitoring the World Economy 1820–1992*, Paris: OECD.

Mauer, Marc, 1997, *Americans behind Bars: US and International Use of Incarceration 1995*, The Sentencing Project (http://www.sentencingproject.org).

Merrill Lynch and Gemini Consulting, 2000, *The World Wealth Report 2000*, Merrill Lynch and Gemini Consulting.

Mitchell, B.R., 1998a, *International Historical Statistics: Europe 1750–1993* (fourth edition), London: Macmillan Reference.

Mitchell, B.R. 1998b, *International Historical Statistics: The Americas 1750–1993* (fourth edition), London: Macmillan Reference.

Mitchell, B.R. 1998c, *International Historical Statistics: Africa, Asia and Oceania 1750–1993* (third edition), London: Macmillan Reference.

Murray, Christopher J.L., and Alan D. Lopez, eds, 1996, *The Global Burden of Disease: A Comprehensive Assessment of Mortality and Disability from Diseases, Injuries, and Risk Factors in 1990 and Projected to 2020*, Cambridge, MA: World Health Organization, World Bank, Harvard School of Public Health.

NLGTF (National Lesbian and Gay Task Force, USA), 1999, website data (www.ngltf.org/downloads/).

OECD (Organization for Economic Cooperation and Development), 1999, Development Assistance Committee, *Development Cooperation Report 1999*, Paris: OECD.

Palestinian Refugee Research Net, 2000, 'Palestinian Refugees: An Overview', article on website (http://www.arts.mcgill.ca/mepp/PRRN/proverview.html).

REIS (Regional Economic Information Service), 1998, website data (http://govinfo.kerr.orst.edu/reis-stateis.html).

SAIRR (South African Institute of Race Relations), 1990, *Race Relations Survey 1989/90*, Johannesburg: SAIRR.

The Sentencing Project, 1999, 'Fact Sheet on Prisons', article on website (http://www.sentencingproject.org/brief/facts-pp.pdf).

SIPRI (Stockholm International Peace Research Institute), 2000, *SIPRI Yearbook 2000: Armaments, Disarmament and International Security*, Oxford: Oxford University Press.

SOPEMI (OECD Continuous Reporting System on Migration), 1999, *Trends in International Migration 1999 edition*, Paris: OECD.

Surfchina, 2000, China statistics, website data (http://www.surfchina.com/).

UBS (Union des Banques Suisses), 1997, *Global Comparison of Wage and Price Levels in 56 Cities around the World*, Zurich: UBS.

UN (United Nations), 1994, Department of Social and Economic Information and Policy Analysis; Population Division, *The Sex and Age Distribution of the World Populations: 1994 revision*, New York: United Nations.

UN (United Nations), 1995, *The Situation of Women in the World 1995: Tendencies and Statistics*, New York: United Nations.

UN (United Nations), 1999, *Demographic Yearbook 1997*, New York: United Nations.

UNCTAD (United Nations Conference on Trade and Development), 1999a, *Handbook of Trade and Development Statistics 1996/1997*, Geneva: United Nations.

UNCTAD (United Nations Conference on Trade and Development), 1999b, *World Investment Report 1999*, Geneva: United Nations.

UNDP (United Nations Development Programme), 1994, *Human Development Report 1994*, New York and Oxford: Oxford University Press for UNDP.

UNDP (United Nations Development Programme), 1995, *Human Development Report 1995*, New York and Oxford: Oxford University Press for UNDP.

UNDP (United Nations Development Programme), 1996, *Human Development Report 1996*, New York and Oxford: Oxford University Press for UNDP.

UNDP (United Nations Development Programme), 1999, *Human Development Report 1999*, New York: UNDP (available from website: http://www.undp.org).

UNDP (United Nations Development Programme), 2000, *Human Development Report 2000*, New York: UNDP (available from website: http://www.undp.org).

UNDP China, 1999, *The China Human Development Report*, Oxford and New York: Oxford University Press.

UNESCAP (United Nations Economic and Social Council for Asia and the Pacific), 2000, 'Population and family planning programme in China by province', website data (http://www.unescap.org/pap/database/chinadata/).

UNFPA (United Nations Population Fund), 1997, *India: Towards Population and Development Goals*, Delhi: Oxford University Press.

UNHCR (United Nations High Commission for Refugees), 1999, statistics on website (http://www.unhcr.ch/statist/).

United Nations Centre for Human Settlements (HABITAT), 1996, *An Urbanizing World: Global Report on Human Settlements*, Oxford: Oxford University Press for Habitat.

United States Bureau of the Census, 1999, *Statistical Abstract of the United States*, 1998 (http://www.census.gov/statab/www).

United States Bureau of the Census, 2000, Statistics on foreign-born population on website (http://www.census.gov/www/documentation/twps0029/).

WHO (World Health Organization), 1999, *World Health Report 1999*, Geneva: WHO.

WHO (World Health Organization), 2000, *World Health Report 2000*, Geneva: WHO.

Wood, Adrian, 1994, *North–South Employment and Inequality*, Oxford: Oxford University Press.

World Bank 1995a, *World Development Report 1995*, Oxford: Oxford University Press.

World Bank 1995b, *Nigeria: Poverty in the Midst of Plenty; The Challenge of Growth with Inclusion*, a World Bank poverty assessment, Washington DC: World Bank.

World Bank 1996, *World Debt Tables 1996*, Washington DC: World Bank.

World Bank 1999a, *Global Development Indicators 1999*, CD-ROM, Washington DC: World Bank.

World Bank 1999b, *Annual Report*, Washington DC: World Bank.

World Bank 2000a, *Global Development Finance 2000*, Washington DC: World Bank.

World Bank 2000b, *Global Development Indicators*, CD-ROM, Washington DC: World Bank.

World Bank 2000c, *Income Poverty – The Latest Global Numbers*, Washington DC: World Bank (http://www.worldbank.org/poverty/data/trends/income.htm).

World Bank 2000d, *Global Development Finance 2000*, Washington DC: World Bank.

World Conservation Monitoring Centre 1999, maps on website (http://www.wcmc.org.uk/forest/data), Cambridge, UK.

WRI (World Resources Institute) 199, *World Resources 1996–97*, New York and Oxford: Oxford University Press.

WRI (World Resources Institute) 1998, *World Resources 1998–99*, New York and Oxford: Oxford University Press

Development studies
titles from Zed Books

Nassau Adams: *Worlds Apart: The North–South Divide and the International System*

A. Adepoju (ed.), *Family, Population and Development in Africa*

Brian C Aldrich and Ravinder S. Sandhu (eds), *Housing the Urban Poor: A Guide to Policy and Practice in the South*

K. Amanor, *The New Frontier: Farmers' Responses to Land Degradation; A West African Study*

Samir Amin, *Capitalism in the Age of Globalization: The Management of Contemporary Society*

Frederique Apffel-Marglin with PRATEC (ed.), *The Spirit of Regeneration: Andean Culture Confronting Western Notions of Development*

Akhtar Badshah, *Our Urban Future: New Paradigms for Equity and Sustainability*

I. Bakker (ed.), *The Strategic Silence: Gender and Economic Policy*

Asoka Bandarage: *Women, Population and Global Crisis: A Political-Economic Analysis*

Tariq Banuri and Frederique Apffel-Marglin (eds), *Who Will Save the Forests? Knowledge, Power and Environmental Destruction*

Jo Beall (ed.), *A City for All: Valuing Difference and Working with Diversity*

Walden Bello, Cunningham and Poh, *A Siamese Tragedy: Development and Disintegration in Modern Thailand*

Walden Bello, Nicola Bullard and Kamal Malhotra (eds), *Global Finance: New Thinking on Regulating Speculative Capital Markets*

Bennett, McPake and Mills (eds), *Private Health Providers in Developing Countries: Serving the Public Interest?*

Robert Biel, *The New Imperialism: Crisis and Contradictions in North/South Relations*

Braidotti, Charkiewicz, Hausler and Wieringa, *Women, the Environment and Sustainable Development*

Rod Burgess, Marisa Carmona and Theo Kolstee (eds), *The Challenge of Sustainable Cities: Neoliberalism and Urban Strategies in Developing Countries*

Steve Burkey, *People First: A Guide to Self-Reliant Participatory Rural Development*

Raff Carmen, *Autonomous Development: An Excursion into Radical Thinking and Practice*

Raff Carmen and Miguel Sobrado (eds), *A Future for the Excluded: Job Creation and Income Generation by the Poor: Clodomir Santos de Morais and the Organization Workshop*

Ricardo Carrere and Larry Lohmann, *Pulping the South: Industrial Tree Plantations and the World Paper Economy*

Fantu Cheru, *The Silent Revolution in Africa: Debt, Democracy and Development*

Andrew Chetley, *A Healthy Business? World Health and the Pharmaceutical Business*

Chevalier and Buckles, *A Land Without Gods: Process Theory, Maldevelopment and the Mexican Nahuas*

Michel Chossudovsky, *The Globalisation Of Poverty: Impacts of IMF and World Bank Reforms*

Marcus Colchester and Larry Lohmann (eds), *The Struggle for Land and the Fate of the Forests*

Bill Cooke and Uma Kothari (eds), *Participation: The New Tyranny?*

C.M. Correa, *Intellectual Property Rights, the WTO and Developing Countries: The TRIPS Agreement and Policy Options*

Emma Crewe and Buzz Harrison, *Whose Development? An Ethnography of Aid*

Peter Custers, *Capital Accumulation and Women's Labour in Asian Economies*

Dalla Costa, *Paying the Price: Women and the Politics of International Economic Strategy*

Bhagirath Lal Das, *An Introduction To The WTO Agreements*

Bhagirath Lal Das, *The WTO Agreements: Deficiencies, Imbalances and Required Changes*

Bhagirath Lal Das, *The World Trade Organization: A Guide to the New Framework for International Trade*

Diplab Dasgupta, *Structural Adjustment, Global Trade And The New Political Economy Of Development*

E. Date-Bah (ed.), *Promoting Gender Equality at Work: Turning Vision into Reality for the 21st Century*

De Koning and Martin (eds), *Participatory Research in Health*

Oswaldo de Rivero, *The Myth of Development: An Emergency Agenda for the Survival of Nations*

Wim Dierckxsens, *The Limits of Capitalism: An Approach to Globalization Without Neoliberalism*

Siddharth Dube, *In the Land of Poverty: Memoirs of an Indian Family, 1947–97*

Mark Duffield, *Global Governance and the New Wars: The Merging of Development and Security*

Graham Dunkley, *The Free Trade Adventure: The WTO, GATT and Globalism, A Critique*

Gustavo Esteva and Madhu Suri Prakash, *Grassroots Post-Modernism: Remaking the Soil of Cultures*

S. Everts, *Gender and Technology: Empowering Women, Engendering Development*

Edesio Fernandes and Ann Varley (eds), *Illegal Cities: Law and Urban Change in Developing Countries*

Jacques B. Gelinas, *Freedom from Debt: The Reappropriation of Development through Financial Self-Reliance*

A.-M. Goetz (ed.), *Getting Institutions Right for Women in Development*

David Gordon and Paul Spicker (eds), *The International Glossary on Poverty*

Denis Goulet, *Development Ethics: A Guide to Theory and Practice*

Wendy Harcourt (ed.), *Feminist Perspectives on Sustainable Development*

Betsy Hartmann and Jim Boyce, *A Quiet Violence: View from a Bangladesh Village*

Bertus Haverkort and Wim Hiemstra (eds), *Food for Thought: Ancient Visions and New Experiments of Rural People*

Heward and Bunwaree (eds), *Gender, Education and Development: Beyond Access to Empowerment*

Susan Holcombe, *Managing to Empower: The Grameen Bank's Experience of Poverty Alleviation*

Terence Hopkins and Immanuel Wallerstein et al: *The Age of Transition: Trajectory of the World-System, 1945-2025*

Rounaq Jahan, *The Elusive Agenda: Mainstreaming Women in Development*

K.S. Jomo (ed.), *Tigers in Trouble: Financial Governance, Liberalisation and the Economic Crises in East Asia*

Michael Kaufman with Harold Dilla Alfonso (eds), *Community Power and Grassroots Democracy: The Transformation of Social Life*

N. Khoury and V. Moghadam (eds), *Gender and Development in the Arab World: Women's Economic Participation*

Koivusalo and Ollila, *Making a Healthy World: Agencies, Actors and Policies in International Health*

Rajni Kothari, *Poverty: Human Consciousness and the Amnesia of Development*

Serge Latouche, *In the Wake of the Affluent Society: An Exploration of Post-Development*

Arthur MacEwan, *Neo-liberalism or Democracy: Economic Strategy, Markets and Alternatives for the 21st century*

John Madeley, *Big Business, Poor Peoples: The Impact of Transnational Corporations on the World's Poor*

Hans-Peter Martin and Harald Schumann, *The Global Trap: Globalization and the Assault on Prosperity and Democracy*

John Martinussen, *Society, State and Market: A Guide to Competing Theories of Development*

Manfred Max-Neef, *From the Outside Looking In: Experiences in 'Barefoot Economics'*

Manfred Max-Neef, *Human-scale Development: Conception, Application and Further Reflections*

John May (ed.), *Poverty and Inequality in South Africa: Meeting the Challenge*

Patrick McCully, *Silenced Rivers: The Ecology and Politics of Large Dams*

Zakes Mda, *When People Play People: Development Communication through Theatre*

Mengisteab and Logan (eds): *Beyond Economic Liberalization: Structural Adjustment and the Alternatives*

John Mihevc, *The Market Tells Them So: The World Bank and Economic Fundamentalism in Africa*

Ronaldo Munck and Denis O'Hearn (eds), *Critical Development Theory: Contributions to a New Paradigm*

Mundigo and Indriso (eds), *Abortion in the Developing World*

K. Nurnberger, *Prosperity, Poverty and Pollution: Managing the Approaching Crisis*

K. Nurnberger, *Beyond Marx and Market: Outcomes of a Century of Economic Experimentation*

Leif Ohlsson (ed.), *Hydro-Politics: Conflicts over Water as a Development Constraint*

John Overton and Regina Scheyvens (eds), *Strategies for Sustainable Development: Experiences from the Pacific*

Helen Pankhurst, *Gender, Development and Identity: An Ethiopian Study*

Govindan Parayil (ed.), *Kerala: The Development Experience: Reflections on Sustainability and Replicability*

James Petras and Henry Veltmeyer, *Globalization: The New Face of Imperialism*

Riccardo Petrella, *The Water Manifesto: Arguments for a World Water Contract*

Pirotte, Husson and Grunewald, *Responding to Emergencies and Fostering Development*

Majid Rahnema with Victoria Bawtree (eds), *The Post-Development Reader*

Gilbert Rist, *A History of Development: From Western Origins to Global Faith*

Kalima Rose, *Where Women are Leaders: The SEWA Movement in India*

Eric Ross, *The Malthus Factor: Poverty, Politics and Population in Capitalist Development*

Wolfgang Sachs (ed.), *The Development Dictionary: A Guide to Knowledge as Power*

Wolfgang Sachs, *Planet Dialectics: Explorations in Environment and Development*

Saral Sarkar, *Eco-Socialism or Eco-Capitalism? A Critical Analysi of Humanity's Fundamental Choices*

Frans Schuurman (ed.), *Beyond the Impasse: New Directions in Development Theory*

Ian Scoones et al, *Hazards and Opportunities: Farming Livelihoods in Dryland Africa*

Vandana Shiva, *The Violence of the Green Revolution: Third World Agriculture, Ecology and Politics*

Vandana Shiva, *Staying Alive: Women, Ecology and Development*

Harry Shutt, *The Trouble with Capitalism: An Enquiry into the Causes of Global Economic Failure*

Kavaljit Singh, *The Globalisation of Finance: A Citizen's Guide*

Kavaljit Singh, *Taming Global Financial Flows: Challenges and Alternatives in the Era of Financial Globalization*

S. Sittirak, *Daughters of Development: Women in a Changing Environment*

Patricia Smyke, *Women and Health*

Rehman Sobhan, *Agrarian Reform and Social Transformation: Preconditions for Development*

J. Stein, *Empowerment and Women's Health: Theory, Methods and Practice*

Stiefel and Wolfe (eds), *A Voice for the Excluded: Popular Participation in Development*

Mohamed Suliman (ed.), *Ecology, Politics and Violent Conflict*

Bob Sutcliffe, *100 Ways of Seeing an Unequal World*

Ian Tellam (ed.), *Fuel for Change: World Bank Energy Policy, Rhetoric versus Reality*

Thompson and Thompson, *The Baobab and the Mango Tree: Lessons about Development — African and Asian Contrasts*

Susanne Thorbek, *Gender and Slum Culture in Urban Asia*

Richard Tomlinson, *Urban Development Planning: Lessons for the Economic Reconstruction of South African Cities*

Meredith Turshen, *The Politics of Public Health*

Oscar Ugarteche, *The False Dilemma: Globalization – Opportunity or Threat?*

Van Veldhuizen, Waters-Bayer and de Zeeuw, *Developing Technology with Farmers: A Trainer's Guide for Participatory Learning*

Jean Vickers, *Women and the World Economic Crisis*

Visvanathan, Duggan, Nisonoff and Wiegersma (eds), *The Women, Gender and Development Reader*

Petra Weyland and Ayse Oncu (eds), *Space, Culture and Power: New Identities in Globalizing Cities*

Paul Wolvekamp (ed.), *Forests for the Future: Local Strategies for Forest Protection, Economic Welfare and Social Justice*

David Woodward, *The Next Crisis? Foreign Direct and Equity Investment in Developing Countries*

Mathew Zachariah and R. Sooryamoorthy, *Science in Participartory Development: Achievements and Dilemmas of a Development Movement; The Case of Kerala*

For full details of this list and Zed's other subject and general catalogues, please write to: The Marketing Department, Zed Books, 7 Cynthia Street, London N1 9JF, UK or email Sales@zedbooks.demon.co.uk

Visit our website at: http://www.zedbooks.demon.co.uk